Urban Pest Control

A Practitioner's Guide

Urban Pest Control

A Practitioner's Guide

Partho Dhang, PhD

Entomologist and Pesticide Consultant
Manila, Philippines

CABI is a trading name of CAB International

CABI	CABI
Nosworthy Way	745 Atlantic Avenue
Wallingford	8th Floor
Oxfordshire OX10 8DE	Boston, MA 02111
UK	USA

Tel: +44 (0)1491 832111
Fax: +44 (0)1491 833508
E-mail: info@cabi.org
Website: www.cabi.org

Tel: +1 (617)682-9015
E-mail: cabi-nao@cabi.org

A catalogue record for this book is available from the British Library, London, UK.

Library of Congress Cataloging-in-Publication Data

Names: Dhang, Partho, author.
Title: Urban pest control : a practioner's guide / Partho Dhang.
Description: Boston, MA : CABI, [2018] | Includes bibliographical references and index.
Identifiers: LCCN 2018002403 (print) | LCCN 2018003555 (ebook) | ISBN
    9781786395153 (ePDF) | ISBN 9781786395160 (ePub) | ISBN 9781786395146 (alk. paper)
Subjects: LCSH: Urban pests--Control. | Insect pests--Control
Classification: LCC SB603.3 (ebook) | LCC SB603.3 .D43 2018 (print) | DDC 363.7/8--dc23
LC record available at https://lccn.loc.gov/2018002403

ISBN-13: 978 1 78639 514 6 (hbk)
          978 1 78639 515 3 (PDF)
          978 1 78639 516 0 (ePub)

Commissioning editor: Rachael Russell
Editorial assistant: Emma McCann
Production editor: Tim Kapp

Typeset by SPi, Pondicherry, India
Printed and bound in the UK by CPI Group (UK) Ltd, Croydon, CR0 4YY
First printed 2018
Reprinted 2019

Dedicated to
my office staff

Contents

Acknowledgements

I decided to take on the arduous task of putting this book together after receiving innumerable requests from pest control practitioners for an affordable and simplified guide. I would like to thank them all for the confidence they have shown in me. I must admit I enjoyed the work thoroughly.

This book would not have been possible without the people who supported my endeavour from the start. Friends, researchers, business associates, colleagues and office staff around the world knowingly or unknowingly contributed invaluably to my knowledge of entomology, and even more so to making a business of it. I have to name a few people here, without whom this work could not be called complete: M.C. Muralirangan, K.P. Sanjayan, V. Mahalingam, James E. Baker, Peter Prangley, Jacque Louw, Muhammed Kufrevi, David Nimocks, Steven Broadbent, Pawel Swietoslawski, David Liszka, Ivan Galev, the staff at CAB International, Ras Patel and his team at Research Information Limited are names that I do not want to miss out.

Sincere thanks are due to all my diligent office staff members, who manned the helm, allowing me to take time off and travel to various conferences, workshops and business events. Without their presence I wouldn't have travelled and learnt as much, or had the courage to write this book.

Finally, I would like to acknowledge researchers and scientists, in particular Dr Ken Walker, senior curator at Victoria Museum Australia, who gave permission to freely use their valuable scientific photographs. This gesture went a long way towards subsidizing the cost of this book. Lastly, thanks to Kitty Hernandez for help with illustrations.

Preface

The business of setting up and running a pest control company involves multi-level skills in biology, chemistry, architecture, engineering, sales, logistics, law and accounting, not to mention a generous dose of common sense – I say this because the primary business involves interaction with people.

This book is an endeavour to bring all of these activities together, with an emphasis on the pest organisms that form the core of the business. The organisms discussed in this book mainly consist of those invertebrates and vertebrates that commonly infest man-made buildings, termed 'structures'. These organisms termed 'pests' or 'urban pests' are mostly resident, living indoors for many generations. Also included are the non-resident pests: those invading structures for food and shelter.

Irrespective of their classification as either resident or non-resident, it is important to know that pests are primarily attracted to structures by human activities, such as their behaviour and habits. Consequently, it is emphasized that pest control can be achieved by modifying human behaviour, changing human habits and improving or altering human living conditions. Secondary intervention methods termed 'pest control', such as use of chemicals, should be considered only when there is a failure to achieve the necessary life style changes.

In addition to human life style, faulty or unmindful constructions are also responsible for attracting and harbouring pests in a structure. This is mostly overcome by modifying or correcting engineering designs.

The role of a modern-day pest controller, termed a 'practitioner', is more in equipping him- or herself with knowledge, understanding and monitoring the human occupants, and checking structures to maintain a pest-free situation. Their actions to intervene with pest control practices such as the use of chemicals and pest control tools become necessary only when the above are not possible or face limitations.

Finally, I would like to mention that this book shares my personal experiences, and if by any chance it shows biases or preconceptions, this should be considered as purely unintentional.

Partho Dhang, PhD
Manila
15 October 2017

1 Understanding the Business of Controlling Pests

1.1 Introduction

Urban pests are common all over the world. These include cockroaches, flies, mosquitoes, bed bugs, ticks, fleas, ants, termites, rodents and a few more. These pests thrive in dark, warm and moist conditions in structures, particularly in places where there is food, warmth and places to hide. Moreover, a number of human activities and habits such as living in homes with insufficient ventilation, creating clutter, poor lighting, temperature control, poor recycling of rubbish, improper composting methods, poor water storage and use of wood in construction attract pests. Also, community and public areas in cities such as parks, recreation centres, wastelands, rivers, canals, sewer drains, stormwater drains, dump sites, flea markets and recycling plants often serve as ideal breeding grounds and habitats for pests.

- *A city is never free of pests, and urban pests are among the prime sources of damage and many human illnesses and injuries.*

Urban pests are the leading causes of illnesses due to allergies, bites, food contamination and phobia. They also harm humans by causing significant damage to property and structure. Consequently, the pest control industry has flourished because of the human need to eliminate them, generating billions of dollars annually. Today we have global pest control companies such as Rentokil, Terminix and Orkin, in addition to many local ones, who have synonymized their brands with the very act of pest control.

1.2 Trading Pest Control Business

Pest control is becoming a necessity for humans (Figs 1.1–1.4). The sight of a pest triggers various types of negative behaviour, such as anger and disgust, and the choice to use a toxic chemical spray is almost involuntary. This human behaviour has made pest control a burgeoning business area. However, the degree of trade is dependent on the nature of the service the practitioner is offering and the environment where the service is required. An occasional trail of ants in the home may be merely a nuisance; in contrast a single ant in a hospital can have serious consequences. The tolerance for pest infestations varies from situation to situation. Institutional kitchens, health care facilities and critical manufacturing facilities demand detailed and careful design and planning to exclude pests compared to homes or shopping centres.

- *Pest control is becoming a necessity for humans. The sight of a pest triggers various types of negative behaviour, such as anger and disgust, and the choice to use a toxic chemical spray is almost involuntary.*

The information disparity with regards to pests, between the client (such as a homeowner) and the pest controller has been utilized to make pest control services a successful business. Practitioners often help to bridge the knowledge gap by providing information, printed literature, client references and web reviews. Practitioners often introduce the products they use in order to gain the confidence of the client. In addition, manufacturers continuously update their product pages with research and findings to help clients gather reassuring information. Another way of getting around this information asymmetry and making pest control businesses more profitable is to offer warranties to gain the client's trust. Pest control companies often provide various types of service and product warranties. Such warranties can be in many forms, such as retreatment without incurring extra cost in the case of the pest resurfacing. Practitioners have even experimented with offers to replace damaged property to increase their business.

- *The information disparity with regard to pests and pesticides between the client and practitioners has been utilized very well to make pest control services a profitable business.*

Fig. 1.1. A museum artefact destroyed by powderpost beetles. (Courtesy of Josielyn Trinidad.)

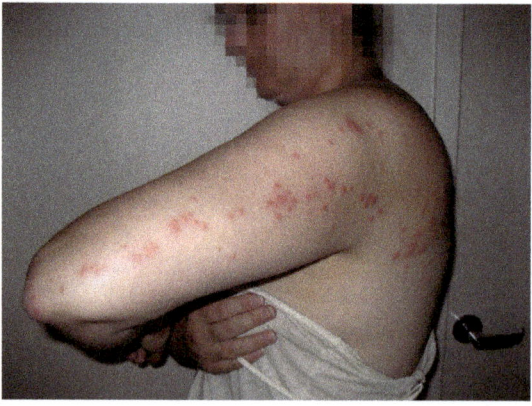

Fig. 1.2. A resident bitten by bed bugs. (Courtesy of Dr Stephen Doggett.)

1.3 Trends in the Pest Control Industry

Current trends indicate that pest control has deviated from a dynamic activity to become more of a ritualistic operation and one that is partly product marketing. Industry-formalized practice is focused on products and their usage. Exhibitions, magazines and journals dedicate vast resources and space to products in comparison to that devoted to knowledge and practices. This forces pest control practitioners to become more product addicted than skill oriented.

- *There is a current trend of overreliance on products, and industry-designed practices have prevented development of skills among practitioners.*

Application of chemicals by sprayers remains the most dominant activity in pest control work. The act is less skilful and less time consuming than other methods of pest control. Sprayers help achieve easy spread of the chemical over a large area in less time. Sprayers cover solid surfaces and water bodies, help treat cracks and crevices and even take care of room space. Their ease of use and the lack of technical skills required to operate sprayers in turn have promoted liquid formulations of pesticides. Formulations are designed to have the broadest possible use, which will cover a variety of insect pests occurring in a habitat. A crack and crevice treatment formulation covers almost all indoor insects that inhabit cracks and crevices. Overall, controlling pests has become a simple act requiring minimal specialized training and knowledge.

Likewise termite management continues to operate under an age-old soil treatment-based business model that has little relevance to actual knowledge. Both consumers and regulatory agencies are largely unaware of the gulf between knowledge and practice. Industry acceptance of a knowledge-based practice model is hindered by a business practice based on insecticide treatment and offers of questionable warranties.

Pest management professionals often choose products based on the treatment cost, and sprayable formulations are often the cheapest. A number of studies, as well as jobs done, have shown the relative ineffectiveness of conventional sprays compared to knowledge-based intervention methods (Dhang, 2014).

- *Pest management professionals who understand the behaviour and biology of their target pest species are more efficient and effective in controlling pests than those who lack this knowledge.*

It is clearly evident that the act of spraying pesticides in an indiscriminate and unplanned fashion has resulted in control failures. This has led to the realization that judicious use of pesticides is needed

Fig. 1.3. A house cabinet destroyed by termites. (Author's photo.)

Fig. 1.4. A textile item damaged by stored product beetle. (Courtesy of Clemson University – USDA Cooperative Extension Slide Series, Bugwood.org.)

to overcome control failures and, to sustain growth, new avenues have to be opened up in terms of application methods. As a consequence of this, dramatic changes in pest control strategies have taken place, which is noticeable globally. Conventional sprayers and indiscriminate sprays have been replaced by precise and targeted delivery systems. Even in cases where spraying is a must, long-lasting residual formulations are used so that the number of sprays is reduced. The advent of baits for a variety of pests has further reduced regular insecticide sprays as residuals. Baits have also, to a large extent, reduced the use of sprayers in the indoor environment.

1.4 Pest Control Tools

Inspection and monitoring are critical tools in pest control. These help make pest control operations sustainable. Both help practitioners determine the right strategy for a job. The knowledge obtained from an inspection is important in making the pest control programme successful. Pest management professionals who understand the behaviour of their target pest species are more efficient and effective at controlling pests than those who lack this knowledge. Knowing common travel routes and typical breeding, hiding and feeding places helps the professional conduct a focused inspection. Instead of wasting time looking where the pest probably is not, time is better spent looking where the pest most likely is. Understanding the harbourage and passage make the work of pest control effective. This helps reduce the cost of treatment and the time spent on the job – each important in making the work sustainable.

- *Even though pest control remains chemical dependent, the nature and mode of action of current chemicals are safer and more target-specific than before. Non-toxic insect growth regulators and physical action products have taken over from contact poisons.*

Use of pest monitoring or intercepting devices is critical in designing a pest control programme.

These devices help detect the presence of a pest, determine the location or active areas of infestation and also indicate the pest population. Each piece of this information is a critical determinant in deciding a suitable pest control strategy. Several active and passive monitors/interceptors are commercially available to assist practitioners.

1.5 Sustainable Pest Control

The market presents various choices of pest control products; however, none permanently eliminate pest problems or makes a structure foolproof against pest attack. This often necessitates the use of multiple methods, including chemical, physical, mechanical and cultural methods, rendering pest control jobs time consuming, laborious and expensive. Thus, long-term pest management has to rely on sustainable methods, such as adoption of integrated pest management (IPM), which combines pest elimination, cost-effectiveness and environmental concerns.

- *Conventional sprayers and indiscriminate sprays have been replaced by innovative and precise targeted delivery systems, such as insecticide baits. Adoption of IPM has further increased the need for continuous monitoring, and lessened the need for treatment.*

Pest control is predominantly an art and requires a dynamic mixture of skill and knowledge. It is generally thought that fundamental design problems in both buildings and landscapes are partly to blame for all pest control failures. It is also a common understanding that pests in the urban environment cannot be permanently eradicated. Thus, there is a need to develop sustainable methodologies to contain pests. These methodologies are key to the development of repeatable practices.

- *The principal revenue for a successful pest control company is generated from providing a continuous monitoring programme rather than mere pest eradication.*

Successful pest control relies on continuous monitoring and maintenance (Fig. 1.5). This realization in recent times has encouraged the invention of a number of pest control products, delivery systems and application methods (Figs 1.6 and 1.7). On the other hand, product development has opened up new methods of application, newer approaches and rationality. The impetus to develop alternative control methods has allowed research on insect baits to take place. Conceptually, baiting systems use major behavioural cues of insect pests that make application methods practical. This appealed to serious pest control practitioners. Baiting soon grew in popularity and acceptance and is widely used in all geographic regions. Insecticide baits remain as the best example of

Fig. 1.5. Providing a continuous monitoring service to clients not only serves to keep a check on pest activity, but also helps generate revenue. (Courtesy of pexels.com.)

Fig. 1.6. Practitioners need to upgrade their knowledge continuously through conferences and workshops to provide the best services to their clients. (Courtesy of Rapid Solution Conference.)

practitioners' approaches to pest management taking a sustainable direction.

- *The role of a modern-day pest controller is more in equipping him- or herself with knowledge and understanding of structures and monitoring human occupants to maintain a pest-free situation. Intervention with pest control chemicals and tools becomes necessary only when the above are not possible or face limitations.*

The success of sustainable pest control remains dependent on understanding the nature of the infested habitat and pest behaviour. In an example with termite management, a retrospective analysis showed that structures with surrounding landscape, owned by the middle classes and constructed by developers have significantly more infestation than other types of structures (Dhang, 2011). Such intrinsic information on infestation patterns is useful in designing sustainable pest management strategies and programmes.

As an example, foolproofing structures by using sound engineering methods is considered to be the best way to prevent termite entry into structures.

The Australian and Asian markets have seen the use of a number of physical barriers, such as stainless steel mesh, special grades of cement and resin mixtures, and insecticide-impregnated plastic sheets in termite proofing buildings. These products take care of construction gaps, cracks and gaps in and around service penetrations when skilfully installed, thus preventing termite entry through concealed points in a building. The popularity of their use is gaining momentum thanks to the realization of the need to keep chemical insecticides out of some construction sites and the movement towards green building. Thus, using relatively simple design features can substantially reduce long-term pest control costs in buildings and landscapes, while also reducing the health and environmental impacts of pesticide use.

1.6   Green Pest Control

- *Recent trends show indiscriminate use of the terms 'green' and 'organic' in pest control, which may be incorrect representation and consumers need to be made aware of this.*

Fig. 1.7. Practitioners need to improve their skill continuously by attending on-the-job training to provide better services. (Author's photo.)

The colour and the word 'green' are commonly associated with nature, vivacity, life, spring time, freshness, youth, inexperience, hope, safety, permission, etc. The 20th century saw green being associated with environmentalism and environment-related movements. The pest control industry also used 'green' to depict many qualities. However, recent trends show a large increase in the indiscriminate use of 'green' and 'organic' in pest control. The definition differs among practitioners, but generally it refers to the following:

- safety;
- responsibility;
- less use of pesticides; or
- use of an alternative to pesticides.

1.7 Pesticides and Health

Increased urbanization has made pest occurrence a common concern in cities around the world. This phenomenon eventually increased the use of pesticides. The sale of pesticides in supermarkets has further allowed free and indiscriminate usage. Consequently, the chances of mass human exposure to pesticides have also increased in recent years.

- *Indoor application of pesticides, which are regulated by complex risk assessments before and after they are put on the market, does not pose any risk if they are applied by a trained professional as per the label directions.*

Data on the effect of pesticides on human health are mostly generated from occupational settings and dietary exposures. However, very limited data are available on the health effects of indoor exposure to various pest control activities. Seemingly, indoor application of pesticides, which are regulated by complex risk assessments before and after they are put on the market, does not pose a high level of risk if the application of the product and the management of the application take place according to label directions and with proper precautions. However, the presence of vulnerable and unsuspecting members of the population, such as children, has to be taken into consideration every time a treatment is conducted.

2 Household Pests and Their Control – Cockroaches

Cockroaches are the most common urban pests in structures across the world. Worldwide there are around 3500 species of cockroaches, of which only a few species prefer the constant temperature and moist conditions that humans maintain in their homes and working areas. These few species have attained pest status and need to be controlled.

- *Cockroaches are attracted to warm and humid places, they are capable of substrate manipulation, known for hygienic behaviour, food sharing, cannibalism, vibrational communication, kin recognition, trail following and care of the brood.*

Cockroaches are associated with various public health problems due to their association with human waste and their ability to move between filthy areas into homes and commercial establishments. At least 22 species of pathogenic human bacteria, viruses, fungi and protozoans, as well as five species of helminthic worms, have been isolated from field-collected American cockroaches. Cockroaches are also medically proven to be the cause of allergies in children and adults.

All species of cockroaches follow an incomplete life cycle as it goes through egg–nymph–adult stages (Table 2.1). Cockroach nymphs are often confused with other cockroach species as well as with other insects such as bugs and beetles. A professional may be needed to determine the difference at times. A few notable species of pest cockroaches are discussed in the following section.

2.1 German Cockroach (*Blattella germanica*)

The German cockroach is the most common household pest encountered in homes and commercial buildings. It is a resident pest. The pest is attracted to the food humans prepare, eat and throw away. They live in cracks, crevices and dark spots in the building provided by furniture, wall cabinets and numerous household items, which provide the right type of conditions for growth and development.

Description

Adult German cockroaches are 1.1–1.6 cm long. Individuals are brown to dark brown in colour, with two black stripes on the pronotum (the area just behind the insect's head). Females are usually darker and their abdomens are more rounded (Figs 2.1 and 2.2).

Nymphs in the first stage are very dark in colour. In later stages, a pale tan stripe appears down the middle from front to rear. This stripe divides the nymphal markings into two dark, long stripes. The stripes remain as two dark streaks on the adult's pronotum, while the rest of the body is covered by pale brown wings.

2.1.1 Life cycle

Eggs

The egg capsule of the German cockroach is about 5.0–6.5 mm long. Half of it protrudes from the female's abdomen when developed. It is carried in this way for 3 weeks until it is dropped, about 1 day before the eggs are ready to hatch. Each egg capsule contains 30–40 eggs. The female is capable of producing four to eight capsules in her lifetime. The first few capsules will have a full complement of eggs, but subsequent capsules can contain fewer. Female German cockroaches are known to safely deposit mature capsules in hideouts protected from disturbance and exposure.

Nymphs

The eggs hatch roughly a month from the day they are deposited. Nymphs moult six or seven times before reaching the adult stage. Females often have one extra moult compared to males.

Table 2.1. Various life history details of common cockroach species. Notice a significant variation – this is due to the fact that the life cycle of the cockroach is dependent on temperature, humidity and quality of food.

	German	American	Brown-banded	Oriental
Adult body length (cm)	1.1–1.6	3.5–5.0	1.0–1.5	2.5–3.2
Egg case length (max.) (mm)	6.5	9.5	6.5	9.0
Eggs per egg case	30–40	6–28	16–18	15–18
No. of nymphal instars	5–7	7–18	6–8	7–10
Development time (days)	60–120	100–150	90–250	200–600

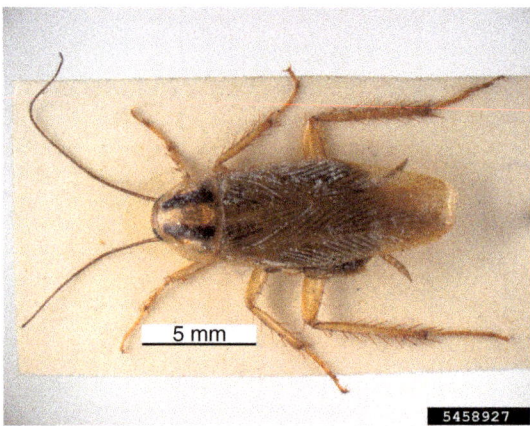

Fig. 2.1. Typical features of an adult German cockroach. (Courtesy of Pest and Diseases Image Library, Bugwood.org.)

Fig. 2.2. German cockroach with distinguishing marks on the pronotum as the principle identifying characteristic. (Courtesy of Pest and Diseases Image Library, Bugwood.org.)

Adults

Adult cockroaches emerge from the final moult with well-developed wings. Emerged adults live in small aggregated groups. The life span of the adult female varies greatly but can be generalized to be between 20 and 30 weeks.

Mating

Females are ready for mating 1 week after becoming adult. Sex pheromones are used to locate a potential mate. After mating females secure a hiding place where they can develop the eggs and the egg capsule.

2.1.2 Control and management of the German cockroach

Identifying hiding places

The most commonly used harbourage sites are under and behind food dispensers, refrigerators, stoves, sinks, dark and stacked up cabinets, etc. In severe infestations cockroaches can be found inside each and every item around, such as electrical outlets, behind wall mounted fixtures, fume hoods, etc. Kitchens and bathrooms provide the most favourable humidity levels, thus becoming the favoured locations.

Inspection

Inspection is the most critical work when it comes to management of German cockroaches. Tools needed for inspection are:

- a good flashlight/torch;
- a rubber handled screwdriver;
- a handy mirror, preferably with extendable handle;

- a magnifying glass;
- glue traps for passive detection;
- glass jar traps baited with fermenting materials such as beer, bread or potatoes.

Search

A typical search should begin in the kitchen, bathroom, storage areas, dining areas, behind and under food processing equipment, and loading and unloading areas. The search should then extend to all adjacent areas. The technician should search all areas, particularly dark, undisturbed or inaccessible ones using the tools listed. Location of live roaches, including nymphs and adults, roach droppings, egg cases and roach smell should be noted.

The use of glue traps and glass jar traps is a common inspection or monitoring method for passive roach detection. These passive traps are left in suspected areas and they are revisited after a few days to detect the presence of a population. Correct trap placement based on foraging habits is a must for increased chance of detection.

Habitat modification

Once German cockroaches have been identified infesting the location, swift action to control the existing population by using recommended insecticides is advised. The next step would be to explain the changes to be made in the habitat that will reduce or eradicate the pest. These recommendations should include how clients can eliminate or restrict material that supports the population.

Recommendations include:

- Keep moisture under check. Repair all leaks to prevent water accumulation.
- Keep food in tightly sealed, roach-resistant containers.
- Keep bags of dry pet food in plastic containers with snap-on lids.
- Store leftover food inside a refrigerator.
- Keep counters, food preparation surfaces, kitchen appliances and floors as clean as possible.
- Periodically, intensively clean kitchen areas, in particular the floor, focusing on areas where grease accumulates, such as drains, vents, ovens and stoves.
- Store food waste and other organic materials in plastic containers with tight-fitting, snap-on lids.

- Put screens on vents, windows and ducts to reduce roach passageways. Caulk around the edges of screens to make a tight barrier.

2.1.3 Recommended insecticide treatment

1. *A crack and crevice type of pesticide application* is preferred for eradicating the existing population. Using a narrow-diameter extension tube attached to a compressed air sprayer nozzle for precision application is recommended. The insecticide used for this application should be a residual insecticide. This not only kills the existing population, but also prevents re-infestation and reoccupation of the cracks and crevices for the next few weeks. Pyrethroid-based chemicals are recommended.

2. *Spot treatment* in non-food areas, in hiding places as well as perimeter spraying covering baseboards, wall voids, floor and wall joints, underneath furniture, equipment, pipes, floor drains and any surface suspected to serve as passage or shelter. Cooking- and food-related items such as utensils and supplies should be moved from the storage places and edges of the shelves treated. Pyrethroid-based chemicals are recommended (Fig. 2.3).

3. *Baiting by using insecticide baits* is a sustainable method for eradication of infestation and maintaining the area as cockroach free. The method is precise, uses very few chemicals and is ideal for sensitive areas. Food areas are best suited to gel bait treatment. Gel baits should be applied once the location of the infestation is determined. This can be determined by ocular observations, feedback from clients or by placement of passive monitors. Gels can be applied either as free drops, streaks or inside gel bait stations. Applied baits should not be contaminated by chemical sprays. Placing adequate numbers of baits around activity areas and harbourage sites is often necessary to achieve the desired results (Fig. 2.4). (More about cockroach baiting is discussed on p. 98.)

4. *Baiting with sprayable forms of insecticide bait* is the most recent technology available to control cockroaches. A food attractant and an insecticide are mixed in a formulation as a sprayable product. Using a conventional sprayer, the product can be quickly applied as spots, making the job cost effective and cheaper. A product containing food bait attractant and 26 gm/l fipronil is now available in many markets around the world. Spot (10–20 ml solution) sprays in and around hideouts and places frequented by cockroaches are recommended. Cockroaches are drawn out and are killed by walking

Fig. 2.3. Spot treatments using a compressed air sprayer allow treatment of cracks and crevices. (Author's photo.)

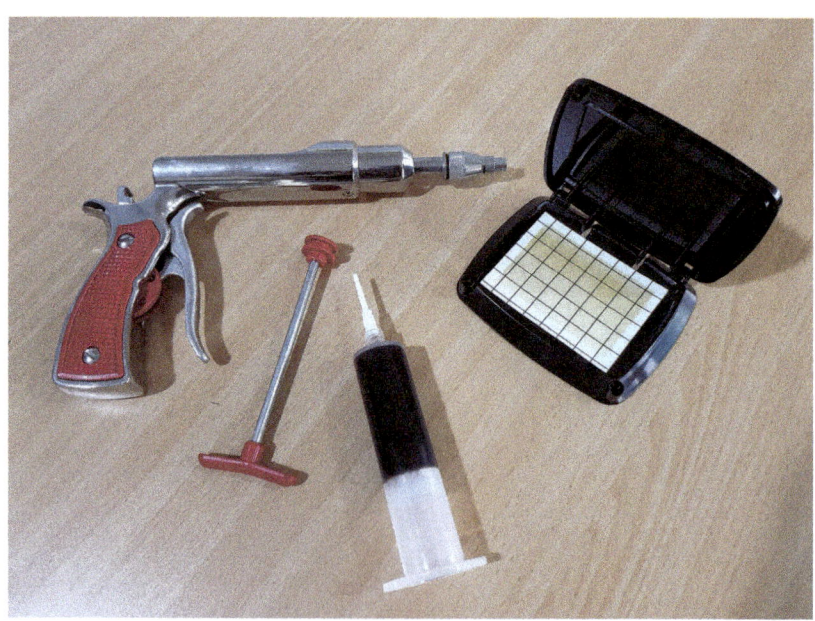

Fig. 2.4. Gel bait, calibrated bait applicator and monitoring stations are usually adequate for managing German cockroaches. (Author's photo.)

over the treated surfaces (Fig. 2.5). (More about this product is discussed on p. 99.)

2.1.4 Monitoring and follow-up visits

It is strongly recommended that, whatever the method of treatment, a follow-up visit in 2 weeks is a must to get the best result. In addition to ocular inspection, cockroach monitors – such as passive monitors – can be placed in strategic locations after a treatment. This helps the technician record cockroach activity on the next visit or helps plan the next course of action. Such information is helpful in understanding the problem over time and in providing clear communication to the clients.

- *It should be noted that, in general, no cockroach infestation, large or small, can be totally eradicated in one single treatment.*

2.2 American Cockroach (*Periplaneta americana*)

The American cockroach is a cosmopolitan pest distributed throughout the world. The species is common in store rooms, cabinets, manholes, drains and practically any dark place in a structure. It is also commonly seen on public transport such as trucks, buses, aeroplanes and ships.

Appearance

An adult American cockroach can grow from 35–50 mm in length. The wings of the male extend slightly beyond the tip of the abdomen, but those of the female do not. This roach is reddish-brown in colour and its pronotum is ringed by an irregular light colour that is almost yellow. Often this margin is bright and wide, darkening towards the centre of the pronotum. In other cases, the lighter margin is barely visible, but it is always present on the rear margin of the pronotum (Fig. 2.6).

Diet

The American cockroach can survive on practically anything. It is an omnivorous and opportunistic feeder. It consumes anything from decaying organic matter to paper, boots, hair, bread, fruit, book bindings, fish, groundnuts, old rice, the soft part on the inside of animal hides, cloth and dead insects.

2.2.1 Life cycle

Eggs

The American cockroach female drops her egg capsules once ready. The capsules are between 5 and 6 mm in size. The egg capsule is dark in colour and dropped in a dark and protected place. The American cockroach can also live outdoors; in such cases egg capsules can be found in moist places. Females produce egg capsules throughout the year. Females can produce from 12–24 capsules in their life time. An average of 14–16 eggs per capsule hatch in 30–50+ days. The numbers indicated above vary depending on climate.

Fig. 2.5. Spot application of sprayable bait using a compressed air sprayer saves time and covers a large area. (Courtesy of ICB Pharma.)

Nymphs

When they first hatch, nymphs are lightly coloured. After their first moult, they attain a reddish-brown colour, resembling adults. They moult from 6–14 times before reaching adulthood. Depending on temperature nymphs can take from 6–20 months to mature (Fig. 2.7).

Adults

Adults commonly live more than 1 year, and up to 22 months. Males and females have a pair of slender, jointed cerci at the tip of the abdomen. The male cockroaches have cerci with 18–19 segments, while the females' cerci have 13–14 segments. The male American cockroaches have a

Fig. 2.6. Typical features of an adult American cockroach. (Courtesy of Daniel R. Suiter, University of Georgia, Bugwood.org.)

Fig. 2.7. Various nymphal stages of American cockroach, often confused with other cockroach species. (Courtesy of Daniel R. Suiter, University of Georgia, Bugwood.org.)

pair of styli between the cerci, while the females do not.

2.2.2 Control and management of the American cockroach

Identifying hiding places

The most used harbourages common to the American cockroach are store rooms, stairwells, around cabinets, drains, manholes, sewers and any area that is dark, damp and undisturbed.

Inspection

Inspection is the most critical work when it comes to control of American cockroaches. Tools are needed for inspection as indicated under the German cockroach control section. Location of live cockroaches including nymphs and adults, roach droppings, egg cases and roach smell should be noted to detect harbourage.

Habitat modification

Once an American cockroach population has been identified infesting the location, swift action using insecticides to control the existing population is recommended. This should be followed by habitat changes to be made in consultation with the client in places that could assist in reducing or eradicating the pest. These recommendations should include directions on how clients should help restrict incoming materials, which often carry new populations.

- Caulk around plumbing and other penetrations in walls; install wire mesh screens on drains and floor drains; keep drain traps full or capped.
- Remove clutter, and organize stores so they can be inspected regularly.
- Ventilate humid places.
- Vacuum or wipe areas where egg cases would be left for incubation.

2.2.3 Recommended insecticide treatment

1. *Residual insecticide and spot spray* on all types of surfaces, such as concrete floors, bricks, stones, soil, etc., which serve as pathways for cockroach movement. Apply insecticide in cracks and crevices, and hideouts, as recommended for German cockroach treatment.

2. *Space sprays or thermal fogging* using non-residual insecticides to quickly reduce large populations indoors, as well as covered areas outdoors. Places such as manholes, concealed drains, ceiling voids, stairwells and elevator shafts are intermediate areas in structures that are perfect for space treatments. Pyrethroid-based chemicals are recommended (Fig. 2.8).

Fig. 2.8. A public manhole being treated with a thermal fogger against American cockroaches. (Courtesy of Biosav.)

2.3 Brown-banded Cockroach (*Supella longipalpa*)

The brown-banded cockroach is generally not as widespread and common as the German cockroach or American cockroach. However, if introduced into a structure they can find favourable harbourage, such as the ideal temperature and moisture levels to build up a population.

Appearance

Adult brown-banded cockroaches can grow between 10 and 15 mm in length. Both sexes have a light band behind the pronotum at the base of the wings, and another full or partial band about one-third of the way back from the pronotum. The pronotum is dark brown with very light/colourless side margins and does not have two stripes as in the German cockroach. Nymphs are dark with two very light bands separated by a dark band just behind the pronotum (Figs 2.9 and 2.10).

2.3.1 Life cycle

Eggs

The brown-banded cockroach female forms an egg capsule, carries it for less than 2 days and then glues it to an object in the harbourage site. The capsule is very small, about 3 mm long. It is oval and a light tan to brown in colour. The female usually glues these in clumps underneath furniture, behind kitchen cabinet drawers, and in corners inside cabinets and cabinet frames. These capsules hatch in about 30–50 days. A female may deposit 14 egg cases in her lifetime; 13–18 nymphs can hatch from one egg case.

Nymphs

Nymphs moult six to eight times to become adult. The duration of nymphal development is around 3–6 months in Asian conditions.

2.3.2 Control and management of brown-banded cockroach

Identifying hiding places

Brown-banded cockroaches prefer warm and dry locations, such as near refrigerator motor housing, on the upper walls of cabinets, and inside pantries, closets, dressers and furniture in general. They can

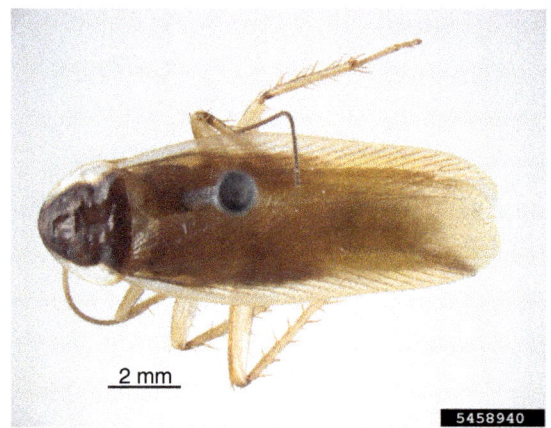

Fig. 2.9. Typical features of an adult brown-banded cockroach. (Courtesy of Pest and Diseases Image Library, Bugwood.org.)

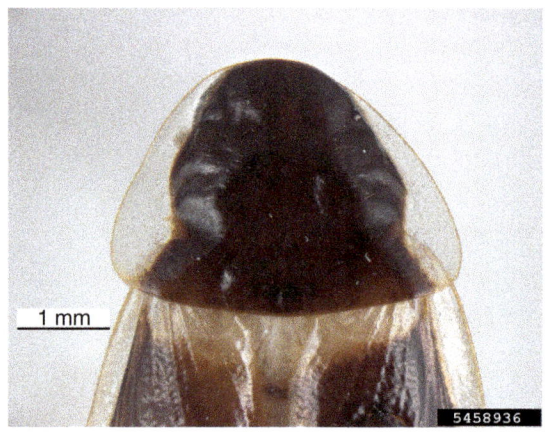

Fig. 2.10. Brown-banded cockroach. The pronotum is dark brown with very light side margins, without the two stripes of the German cockroach. (Courtesy of Pest and Diseases Image Library, Bugwood.org.)

also be found behind picture frames and beneath tables and sofas, and inside clocks, radios, light switch plates and door frames. It is common to find them hiding nearer the ceiling than the floor and away from water sources.

2.3.3 Recommended insecticide treatment

1. *Dry-flowable bait formulation* has deeper penetrating properties into cracks and is useful in controlling the species. Generally populations of brown-banded cockroaches are isolated and small.

Baits need to be applied only to those areas where cockroaches are harboured.

2. *Spot treatment* using a narrow-diameter extension tube and a residual formulation in all infested cracks, crevices and hideouts is recommended. Areas such as under furniture, drawers and sinks, and around pipes and high cabinets need to be covered. Pyrethroid-based chemicals are recommended.

3. *Caulking cracks and spaces* as well as regularly vacuuming potential hideouts is recommended.

2.4 Oriental Cockroach (*Blatta orientalis*)

The Oriental cockroach is generally not as widespread and common as the German or American cockroaches; however, isolated populations are common. The Oriental cockroach is often confused with a beetle due to its general appearance.

Appearance

The adults of the Oriental cockroach are very different in appearance to other cockroaches. The smaller adult male Oriental cockroach is 25 mm in length, and can be identified by the presence of three-quarter-length wings, leaving the last few abdominal segments exposed. The larger adult female Oriental cockroach measures 32 mm in length and totally lacks wings, having only large wing pads that cover the first couple of segments of the body. Obviously, neither the male nor female is capable of any flight (Figs 2.11 and 2.12).

2.4.1 Life cycle

Eggs

Egg capsules of the Oriental cockroach may appear dark brown or reddish in colour and are almost 8–10 mm in length. Each egg case, which can hold approximately 16 eggs, is dropped by the female in protected areas, almost 1 day after it is produced. Eggs hatch in 1–2 months.

Nymphs

Nymphs moult seven to ten times and are reddish-brown to black in colour, except in the first stage, when they are pale tan. The older brown Oriental cockroach nymphs are very difficult to distinguish

Fig. 2.11. Typical features of an adult Oriental cockroach, which can be identified by the presence of three-quarter-length wings, leaving the last few abdominal segments exposed. Females lack wings totally. (Courtesy of Pest and Diseases Image Library, Bugwood.org.)

Fig. 2.12. Oriental cockroach pronotum without any markings. (Courtesy of Pest and Diseases Image Library, Bugwood.org.)

from American cockroach nymphs. The best way is to capture them alive and rear them to adulthood, which might take a week.

Behaviour

Oriental cockroaches prefer cool, damp locations, so they are typically located in basements, cellars, fresh food stores, etc. Oriental cockroaches mostly crawl around service ducts, toilets, bathtubs, sinks, radiators and pipes, and can be located easily when present.

2.4.2 Control and management of the Oriental cockroach

Identifying hiding places

The Oriental cockroach is primarily an outdoor species, well adapted for surviving in the natural environment, such as in gardens and planter boxes. Most outdoor populations can be found living beneath the mulch in landscape beds, in leaf litter, beneath stones or debris outdoors. In structures they can thrive in the voids or openings beneath porches, in wall voids and crawlspaces, storm drains and sewers. Oriental cockroaches are known for their preference for feeding on garbage, filth or material that has begun to decay. These cockroaches are very much dependent upon moisture.

Inspection

Search areas of high humidity and hiding places.

2.4.3 Recommended insecticide application

Many of the same insecticide applications and methods used to reduce populations of other cockroaches will work for the Oriental cockroach. Particular attention must be paid to the type or class of insecticides when applied in outdoor areas. The insecticides must be non-phytotoxic. Pyrethroid-based chemicals are recommended.

2.5 Australian Cockroach (*Periplaneta australasiae*)

This species is a rare and occasional pest, with a very close resemblance to the American cockroach. It can be called an occasional pest.

Appearance

The Australian cockroach looks similar to the American cockroach, but is slightly shorter and somewhat oval in general shape. Australian cockroach adults have conspicuous light yellow margins on the pronotum. The reddish-brown base colour is slightly darker, and the outside edges of the wings just behind the pronotum are light yellow, sometimes nearly white. Nymphs are brown, but have yellow streaking across each thoracic (middle section) and abdominal segment (Figs 2.13 and 2.14).

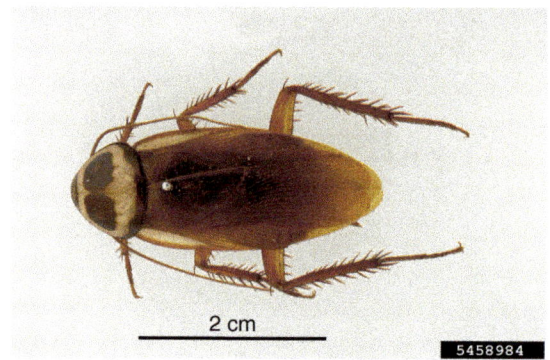

Fig. 2.13. Typical features of an adult Australian cockroach. (Courtesy of Pest and Diseases Image Library, Bugwood.org.)

Fig. 2.14. Australian cockroach adults have conspicuous light yellow margins on the pronotum. (Courtesy of Pest and Diseases Image Library, Bugwood.org.)

2.5.1 Behaviour and habitat

The Australian cockroach is more commonly introduced in trees, potted and ornamental plants used inside shopping malls and commercial buildings. It burrows into soil and is not easily detected. The Australian cockroach can build up in large numbers inside buildings with the presence of high humidity.

2.5.2 Control and management of the Australian cockroach

Inspection

Inspect thoroughly areas such as gardens and planter boxes, in which they commonly burrow. It is best to inspect in darkness with the help of black light for easy detection, as they are often active during night-time hours.

2.5.3 Recommended insecticide application

Bait stations such as glue boards and bait granules containing insecticide can be placed in and around plants to trap and kill the infesting population. Conventional chemical spray is not advisable as the population is always hidden and isolated.

2.6 Smoky Brown Cockroach (*Periplaneta fuliginosa*)

The smoky brown cockroach has a close resemblance to the American cockroach and looks similar in shape but slightly smaller in size.

Appearance

Adult smoky brown cockroaches are slightly over 25 mm long, and both sexes have wings that are longer than the abdomen. Their very dark brown colour is striking; no light markings appear on the pronotum or wings. Nymphs, like adults, are dark brown. Antennal tips of young nymphs are white, and the base segments of the older nymphs' antennae are white (Figs 2.15 and 2.16).

Behaviour and harborage

The smoky brown cockroach is a plant feeder and occurs in greenhouses and gardens. Though it is mainly an outdoor roach, it is often transported indoors. Populations build up outside homes and enter indoors where they live in gutters, floor drains, and can find their way into attics. This cockroach is dependent on moisture. In areas of high humidity, populations can build up and infest every level in a structure.

2.6.1 Life cycle

Eggs

The egg capsule of the smoky brown cockroach is large and dark brown. The female usually glues it to objects in the harborage. An average of 17 eggs are in each capsule; as many as 24 eggs have been found.

Nymphs

Nymphs hatch within 50 days.

Adults

The life cycle of a smoky brown cockroach is about 1 year. Both sexes are capable of flight.

2.6.2 Control and management of the smoky brown cockroach

Inspection

The best places to locate them are planter boxes, gutters, roof overhangs and attics.

Fig. 2.15. Typical features of an adult smoky brown cockroach. (Courtesy of Pest and Diseases Image Library, Bugwood.org.)

Fig. 2.16. Smoky brown cockroaches have dark brown colouration with no markings on the pronotum or wings. (Courtesy of Pest and Diseases Image Library, Bugwood.org.)

2.6.3 Recommended insecticide application

Bait stations such as glue boards and bait granules containing an insecticide can be placed in and around outdoor sites and planter boxes and other suspected areas to trap and kill the infesting population. Conventional chemical sprays are not advisable as the population is always small, hidden and isolated.

2.7 Asian Cockroach (*Blatella asahinai*)

Appearance

The Asian cockroach looks nearly identical to the German cockroach. However, the two dark bands in the pronotum are more distinct and the wings are slightly longer than the body. In addition, their behaviour is completely different to that of the German cockroach (Figs 2.17 and 2.18).

2.7.1 Behaviour and harbourage

The Asian cockroach is essentially an outdoors cockroach. The cockroach lives outside and builds up populations under fallen leaves and mulch cover. It favours shady, moist areas and builds up rapidly under favourable conditions. Unlike most cockroaches, it is attracted to light, and adults fly to lighted windows, doors, yard lights and parking lot lights at dusk. From these points they often crawl into buildings or fly to indoor room lights.

2.7.2 Control and management of the Asian cockroach

Inspection of gardens, trees and waste areas next to buildings and structures is recommended to locate harbourage. Treatment can be similar to what is recommended for other cockroaches.

Fig 2.17. Typical features of an adult Asian cockroach (Courtesy of Natasha Wright, Cook's Pest Control, Bugwood.org.)

Fig 2.18. Asian cockroach nymph. (Courtesy of Natasha Wright, Cook's Pest Control, Bugwood.org.)

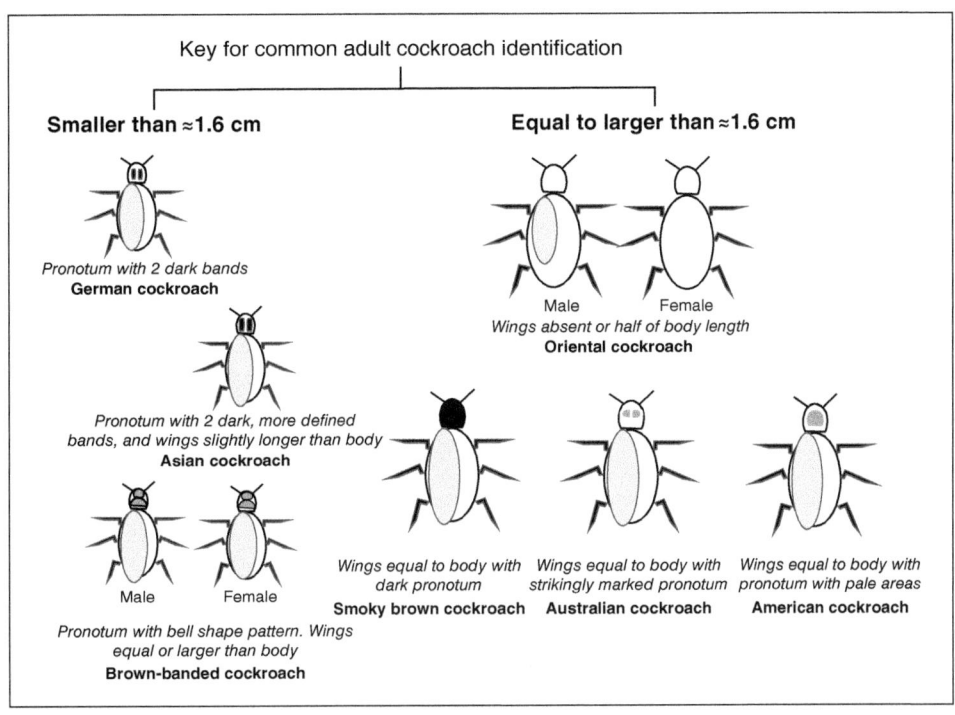

Fig 2.19. Key for common adult cockroach identification.

3 Household Pests and Their Control – Flies

Flies are pests with high public health significance. They are also nuisance pests frequenting garbage, dead animals, decomposing organic matter and farm manure. Their larvae live and develop in organic material. Once emerged, adults fly out to nearby structures and move indoors through open doors and windows, attracted by food flavour, warmth and moisture.

- *Flies have a single pair of wings for flying; their hindwings are modified as halteres, which act as high-speed sensors for rotational movement and allow them to evade approaching objects.*

A few notable species of pest flies are discussed in the following sections.

3.1 Housefly (*Musca domestica*)

The housefly, commonly called '*Musca*' is a non-resident pest. Adult houseflies are capable of transmitting pathogens as a physical carrier (vector) but none of the pathogens undergoes obligatory cyclical development in its body. Adult houseflies grow to 8–12 mm long. The thorax is grey or sometimes even black, with four characteristic longitudinal dark lines on the back. The whole body is covered with hair-like projections. The females are slightly larger than the males, and have a much larger space between their two large compound eyes (Fig. 3.1). Houseflies undergo complete metamorphosis (Fig. 3.2).

3.1.1 Life cycle

Eggs

The eggs of houseflies are 1–1.2 mm long, opal white to cream coloured. Each female can lay an average of 120 eggs per oviposition/laying. Females can lay eggs two to three times a week under favourable conditions.

Larvae

The egg hatches into a larva after between 6 and 12 h at 35°C. Larvae have three larval instars. The larvae are called 'maggots'. The first instar larvae measure approximately 1–3 mm, the second instar 3–5 mm and third instar 5–12 mm in length. The larva has a cylindrical body and a conical anterior tapering and a rounded posterior, with no appendages. The larvae feed on all forms of decomposition products. The larvae try to escape light by burrowing into the food such as manure. The third instar larva is called the pre-pupa. They stop feeding, ready to pupate, and seek drier substrata. The period of development from egg to pupa depends on nutrition and temperature, with a minimum of 3–4 days at 35°C.

Pupae

The pupa is barrel-shaped and brown or dark brown in colour. The duration of the pupal stage depends on humidity and temperature, requiring a minimum of 3–4 days at 35°C and 90% relative humidity (RH). The adult fly emerges from the pupa, breaking open the outer cover and flying out.

Mating and oviposition

Males and females mate the day after emergence in warm climatic conditions. A single mating is enough for females to lay fertilized eggs during her life time. The longevity of females is generally 3–4 weeks. Gravid females are attracted towards the breeding media by its smell. The average numbers of eggs laid by a female per laying could be between 80 and 120. In laboratory conditions nine to ten batches of eggs are usually laid.

Behaviour of adults

Housefly males and females can survive well on water and sugar or a variety of carbohydrates.

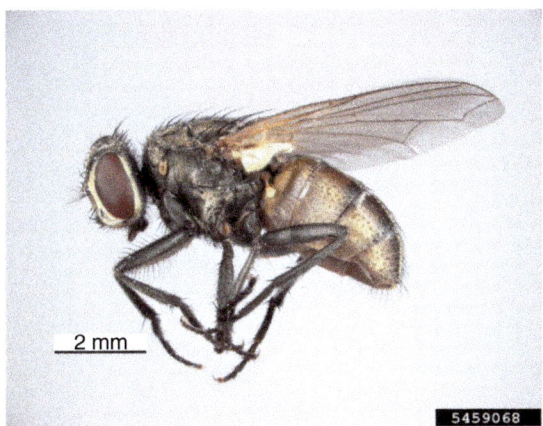

Fig. 3.1. A typical adult housefly. (Courtesy of Pest and Diseases Image Library, Bugwood.org.)

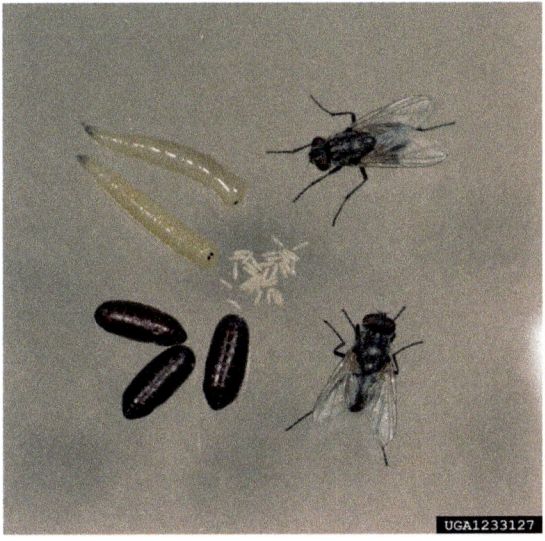

Fig. 3.2. Housefly life cycle depicting a complete metamorphosis with egg, larvae and pupae stages. (Courtesy of Clemson University – USDA Cooperative Extension Slide Series, Bugwood.org.)

However, females require some protein or protein components for the development of eggs. Houseflies feed on all kinds of cooked food, including decomposing organic wastes, fermenting fruits, carcasses, animal manure, etc. Adult flies are only active in daylight or in artificial light, whereas in darkness they crawl slowly and rest mostly on vertical surfaces. Houseflies spend a considerable amount of time shifting from outdoors to indoors seeking the ideal temperature. Houseflies leave their daytime activity and move to special resting sites close to dusk. Preferable resting places are vertical surfaces like wires, cords, cables and walls, and narrow objects away from light.

3.1.2 Control and management of houseflies

Controlling houseflies requires a number of parallel working strategies. These are directed towards adults and larvae separately. Most control measures for adult flies start with treatment of neighbouring breeding sites around the structure. The most common means of fly entry into a structure is through open doors and windows. Modify the door and windows so as to make them close tightly. Install screens or air curtains or install automatic door closers to minimize entry. Replace white security lights inside and outside with yellow lights so flies are less attracted to the building. White light contains a little bit of the ultraviolet (UV) light spectrum, making the light attractive to flies.

Adult houseflies already indoors can be eliminated by insecticide space treatment. This action should be followed by closing or modifying all entry points to the structure and then removing and treating nearby breeding sites. All nearby garbage dumpsters, organic refuse, animal farms and wet soil with organic wastes are areas to be cleaned and treated.

3.1.3 Recommended insecticide application

Depending upon the nature of location and the structure to be treated, a number of methods can be employed for insecticide treatment.

1. *Fly strips or fly ropes/cotton ropes with glue* or glue plus an insecticide can be tied across low-access rooms, such as attics, store rooms, animal farms, dumping areas or slaughter houses. Adult flies, while looking for a resting place, sit on these and are glued. Ropes can be replaced regularly and burnt or appropriately disposed of. This strategy works for a low-level infestation (Fig. 3.3).

2. *Granular scatter fly bait* is useful in eliminating adult flies if the population is small and isolated. Placement of baits and bait stations is critical in this method. The height of bait placement should not exceed 2 m. These work best when placed closer to ground level. This is good for low-level fly population control.

Fig. 3.3. Strings and glue can be tied all around to cut the adult fly population drastically in outdoor areas. (Courtesy of Biosav.)

Dry scatter baits are commercially available. These contain around 0.1–2% of an insecticide in a carrier, which may be plain granulated sugar or sugar plus sand, ground corncobs, sea shells, etc. An additional attractant such as 'muscamone' or 'z-9-tricosene' may be added. The bait should be scattered in thin layers of 60–250 g/100 m² on resting places such as window ledges, etc. At times the baits may need to be made moist to be more attractive (Fig. 3.4). Fipronil-based chemicals are recommended for this.

3. *Paint-on formulations* containing an insecticide and attractant can be used to brush areas where flies generally rest during hot days or night time. This treatment has a residual action that can remain active for a few weeks. However, this is best done in areas that are invisible and hidden, such as animal farms.

4. *Aerosol contact sprays* can be used to knock down adult flies after elimination of breeding sites and exclusion methods are in effect. This is only for small indoor populations.

5. *Spraying residual insecticide* on regular surfaces will kill flies when they are seeking rest. This is the best strategy in all types of infestation situations. Dead flies are knocked off and die on the floor so can be removed by regular housekeeping. Pyrethroid-based chemicals are recommended for this.

6. *Low-volume applications of a non-residual insecticide* using a low-volume applicator is best for treating large outdoor as well as indoor sites. The action is quick and most suitable in outbreak situations. Pyrethroid-based chemicals are recommended (Fig. 3.5).

Fig. 3.4. Fly scatter bait granules are often useful in controlling adult flies in low-intensity indoor infestations. (Author's photo.)

7. *Insect growth regulator or residual insecticide containing formulations* on breeding grounds are most suitable for keeping future populations in check. This prevents larval growth and drastically reduces the adult population. This has to be carried out periodically if the site is a regular breeding location (Fig. 3.6). Pyriproxyfen and cyromazine are recommended for this.

3.1.4 Housefly control by use of traps

Food baited traps

Housefly traps are useful in mass controlling adult flies outdoors in places such as farms. Many types

Fig. 3.5. Low-volume applicator can be used effectively to treat indoor space. (Author's photo.)

of fly traps and fly attractant can be used to make this work. One of the easiest is using melon rinds and molasses in a wire mesh cage. Flies going inside are trapped and desiccate and die. Brown sugar, melon rinds and fish meal can be used as attractants (Figs 3.7 and 3.8).

Water trap using a light source

A simple water trough and a point source of light can be used at times to trap large volumes of houseflies. However, incorporating a black light can dramatically improve the performance (Fig. 3.9).

Housefly control by use of UV light trap

UV light traps are very effective in attracting adult houseflies. Such light traps are very useful only for indoor fly control. This is suitable when a few flies are causing a menace and operations cannot be stopped to undertake a conventional pest control job. The method is discreet, continuous and does

not interfere with regular operations or aesthetics of the area. These traps work round the clock with low maintenance.

- *Housefly exploratory behaviour is very complex and luring them into any contraption is a challenge. Thus, it is a must to have complete knowledge of the environment, which includes a mixture of complex flavours, light, temperature, air draughts and movement, to maximize the efficacy of UV light traps. It is also a must that regular fly control measures should be in place to get the desired results from these traps.*

Insect light traps using a UV light source are devices that evolved out of a need to provide continuous indoor treatment without interference. Light traps were developed for eliminating intruding flying insect pests such as houseflies without chemical intervention or stoppage in operation. UV traps make use of near-ultraviolet (UV-A) 'black light' fluorescent lamps to attract insects, in particular houseflies. To register a high degree of visibility to the flying insect the traps are wall-mounted or suspended from the ceiling. Based on preferred movements and response

Fig. 3.6. Residual insecticide application on organic matter is the best way to keep the larval population in check. (Courtesy of Biosav.)

of flies and night-flying insects, low wall-mounted UV traps have been accepted as the best strategy to maximize catch of both flies and night-flying insects (Figs 3.10 and 3.11).

TRAP LOCATION. Proper location for the installation of these UV traps is of utmost importance not only to gain efficiency, but also to prevent interference with the operation. The first consideration is the placement of traps in areas where people are *not* continuously exposed. Second, the trap location should not hinder its own functioning due to interference from other forms of natural and artificial lighting, should be visible from 8–10 m, not above a height of 2.0–2.5 m from the floor and possibly closer to any exiting air draught. Third, it is important to position them in fly pathways or in 'bottleneck situation areas' in order to trap insects before they reach the critical areas in the facility.

TRAP DESIGNS. The interest in capturing a part of the UV trap market has seen the presence of a number of manufacturers. Though the principal concept remains the same, manufacturers innovate subtle differences to stand out among their competitors. Innovations in aesthetics, open and closed housings, ease of glue board replacement and servicing the traps, easy wall mounting, and choice of material in construction of the housing are the most notable innovations. However, a study is available that investigated the efficiency of various trap designs in trapping flies (Hogsette, 2008). The report shows no significant difference in trap performance between the open and closed housings trap, but the open trap is significantly more efficacious at collecting flies when paired with black light bulbs. The closed trap captures fewer flies than the open trap with the same bulbs, but more than the open trap with the black light blue bulbs. The

Fig. 3.7. Simple backyard housefly trap using an attractant inside a mesh cage trap. (Courtesy of Biosav.)

addition of z-9-tricosene to the glue board had no additional effect on fly catch levels. In spite of these differences in efficacy, black light blue lamps remain in commercial use, simply for the reason that they make the traps less conspicuous.

EFFECT OF LIGHTING ON TRAPS. UV light traps face light coming from other sources, such as room lighting, indirect sunlight, reflected light from objects, etc. Consequently, the efficiency of the traps will be reduced. Regular incandescent and fluorescent lighting have been shown to have no effect on trap performance, if not placed in a position where they attract additional insects to the protected area. However, mercury vapour lamps, which emit significant amounts of UV, can reduce the effectiveness of light traps. The use of sodium vapour and lower wattage lighting will increase light trap effectiveness. LED lights may even prove to be superior since LEDs can be selectively chosen for output on a nanometre scale.

EFFECT OF TEMPERATURE ON TRAP EFFICIENCY. Ambient weather conditions such as air temperature

may influence the attraction of houseflies to colour. A study (Diclaro *et al.*, 2012) has reported that blue may be attractive to flies in general because it is perceived as shaded, cooler resting areas. In addition, adding black lines to the blue targets could make the target more attractive to houseflies perhaps by adding a stimulus to satisfy the scototaxis tendency of houseflies. The houseflies may perceive the blue visual target as a potential resting area and the black lines as cracks or crevices that can be used as harbourage. The same authors suggested that there is a possibility that under cooler conditions, the blue targets would not be as attractive as under relatively warm conditions.

TRAPS AND HUMAN EXPOSURE. Facilities and amenity industries running continuous operations may be required to give clarification regarding the safety of the emitted lights for human occupants in the vicinity. This is an important aspect as people within the area of UV traps may be exposed to reflected and scattered UV-A from the lamp continuously. A variety of UV lights have been used in studies to assess potential exposures. The authors

Fig. 3.8. Mass capture of houseflies is one useful method to reduce the immediate nuisance. (Courtesy of Biosav.)

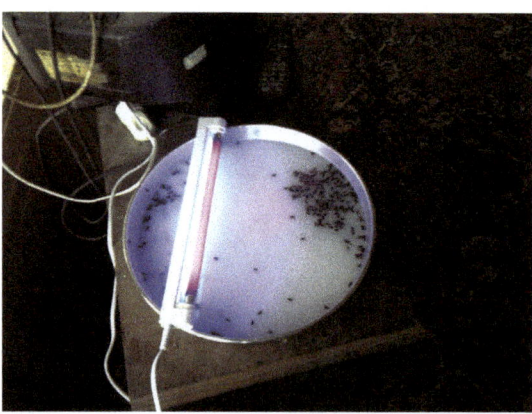

Fig. 3.9. In some situations indoor water traps with a black light can cut the fly population. (Courtesy of Biosav.)

showed that UV-A radiation exposures during any reasonably foreseeable worst-case daily exposure condition would not exceed the currently recommended occupational exposure limits (threshold limit values) for UV-A (in the range of 315–400 nm) for both eyes and skin (Sliney *et al.*, 2016). This observation rules out the harmful effects of UV traps, which at times is a concern raised by clients.

INSTALLING A LIGHT TRAP

1. The height or distance from the floor should ideally be 1.2–1.8 m.
2. The trap should not face sunlight or other light sources.
3. The installation of the trap should follow an outward moving air draught if available.
4. Location of the trap should be away from fume hoods, chimneys, air-conditioning units, doors and windows, and high human activity.
5. The intensity and quality of UV emitted from the bulbs are most critical when determining the efficacy of the trap. It is UV-A that is most attractive to flies and a UV-A measuring device must be used periodically to monitor the intensity of light.

3.1.5 Fly control in special locations

Poultry and animal farms

Flies can be killed directly by insecticides; however, they should preferably be controlled by improving environmental sanitation and hygiene. This approach provides long-lasting results, is more cost-effective and usually has other benefits.

Type of flooring: Solid concrete floors with drains, which can be washed and cleaned daily, must be used on all types of farms. Fans and driers should be used to dry the droppings as much as possible. Leaking water pipes should be repaired. All types of waste and droppings should be removed and the floors should be flushed at frequent intervals (Fig. 3.12).

Converting the droppings into slurry: Converting the droppings or dung quickly into a slurry with water prevents flies from breeding. Female flies will not lay eggs in a water-rich slurry or even if eggs are laid, the eggs will drown. For this a ratio of 50% or more water to solid waste is needed. Insecticide with insect growth regulator (IGR) may be sprayed into the slurry to prevent larval growth.

Drying the droppings: Droppings should be stacked to reduce the surface area for fly breeding. They should be covered with plastic sheets or other fly-proof material. This prevents egg laying and kills larvae and pupae, as the heat produced in the

Fig. 3.10. UV traps work best when the height, location and competing light are considered during installation. (Author's photo.)

Fig. 3.11. UV light traps using glue boards work well in food areas. (Author's photo.)

composting process can no longer escape. It is preferable to stack the dung on a concrete base, surrounded by gutters to prevent the migration of larvae to pupate in soil around the heap. In hot climates, dung may be spread on the ground and dried before the flies have time to develop.

Garbage and other organic refuse

This breeding medium can be eliminated by proper collection, storage, transportation and disposal. In the absence of a system for collection and transportation, garbage can be burnt or disposed of in a specially dug pit. At least once a week the garbage in the pit has to be covered with a fresh layer of soil to stop breeding by flies.

Flies are likely to breed in garbage containers even if they are tightly closed. In a warm climate the larvae may leave the containers for pupation after only 3–4 days. In such places, garbage has to be collected at least twice a week. In temperate climates once a week is sufficient. When emptying the container it is important to remove any residue left in the bottom.

Fig. 3.12. Quickly drying animal droppings by proper aeration and over a soil flooring helps control fly breeding. (Author's photo.)

Garbage should be regularly transported to refuse dumps, to reduce breeding. It is necessary to compact the refuse and cover it daily with a solid layer of soil. Such dumps should be at least several kilometres away from residential areas.

Food processing industries

Flies are attracted by the odour emanating from breeding sites. In addition they are attracted by products such as fishmeal and bonemeal, molasses and malt from breweries, milk and sweet-smelling fruits. Attraction to waste can be prevented by cleanliness, the removal of waste and its storage under cover. Industries using attractive products can install special exhausts for odours. The exhausts can open vertically, upward of 10 m to prevent flies flying in. Physical barriers such as screens, curtains or UV

lights should be used to keep isolated flies in check. Lights used in the building should be monochromatic or neon vapour lamps, as flies are attracted to white light.

Homes and residential places

Insecticides can bring about quick control in residential places such as homes and hotels. Control with insecticides should be undertaken only for a short period when absolutely necessary.

Treating resting sites: The idea of providing toxic resting sites for flies is based on the observation that houseflies prefer to rest at night or during the hot part of the day on vertical surfaces such as edges, hanging strings, wires, cords, ropes, etc. Materials such as ropes, strings, bed nets, curtains, cotton cords, cloth or gauze bands and strong paper strips that can be impregnated with insecticide are useful. This strategy will be effective for many weeks in both tropical and temperate areas. This method is cheap and has long residual action. However, it does not work in rooms with an air draught under the ceiling, which is the case in many ventilated rooms. This technique may look untidy but it is very effective. The surfaces can be periodically removed and replaced.

Dry scatter bait: Dry scatter baits are commercially available and are most useful in low-level infestation.

3.2 Minor Flies as Pests

3.2.1 Blowflies

Blowflies (the family Calliphoridae) are about 15 mm long. Most common in this group are bottleflies, characterized by a thorax and abdomen which are shiny black, metallic green, blue or bronze (Fig. 3.13). They live on dead animals, meat scraps in garbage and wet-mixed garbage. The female blowfly typically lays her eggs on the body of a recently killed animal. The eggs hatch quickly and the maggots then feed on the decaying tissues. In warm weather, some species can complete their larval growth within a week. They then burrow into the soil and pupate, to emerge later as adult flies. Blowflies play an essential role in nature by decomposing dead tissue. The cluster fly species of blowfly is an exception: its larvae prey on earthworms.

Fig 3.13. Typical features of an adult bluebottle fly. (Courtesy of Pest and Diseases Image Library, Bugwood.org.)

Controlling them is basically achieved by using exclusion methods, removal of dead decomposing material and cage traps baited with meat bait.

3.2.2 Drain flies

Drain flies or moth flies are about 3 mm long. Their dark grey or black colour comes from tiny hairs that cover the wings, which are held in roof-like fashion over the body. Moth flies have long, drooping antennae (Fig. 3.14). Larvae live in the gelatinous material in sink drain traps and sewers. Where sinks regularly overflow, these flies build up in the overflow pipe. When drain traps of sinks, commodes and floor drains dry out, large numbers can enter dwellings from the sewer.

In sewage treatment plants, drain flies feed on the gelatinous material that collects on stones in trickling filter beds. Over time, however, cast skins from these filter flies can slow down water drainage.

Control can be achieved by cleaning the drain traps mechanically or with drain cleaners. Without larval control, adults will constantly emerge. The filter bed should be cleaned by reverse- or back-flushing. Granular baits can be used in such locations.

DRAIN FLY
(Diptera Psychodidae)

UGA1252025

Fig. 3.14. Typical feature of an adult drain fly. (Courtesy of Joseph Berger, Bugwood.org.)

3.2.3 Fruit flies

Fruit flies are attracted to nearly any material that is fermented by yeast. These small flies commonly have bright red eyes. The head and thorax are yellowish to brown, and the abdomen is light brown to dark with yellow bands (Fig. 3.15).

In a common fruit fly infestation, flies are attracted to the sweet odour of fermentation in ripe fruit, such as bananas; they lay their eggs in the cracks of the peel. Fruit fly larvae hatch, then feed on yeast cells in the fruit. The life cycle can be completed in 1 week.

Emerged adults are attracted to lights, but egg-laying females will not leave fermenting materials. Fruits, vegetables, beer, fermenting water from refrigerators, humidifiers, sink drains, sour mops and rags, and fermenting pet food are examples of fermenting material.

Infestations are common in breweries, restaurants, canneries, hospitals and homes. Controlling them can be achieved by removing the source, trapping them with attractant baits and misting in cases of severe infestation.

3.2.4 Midges

Midges look very much like mosquitoes and female midges are capable of biting (Fig. 3.16). Midge larvae live in water, especially in quiet, still water. Adult midges are often a food source for spiders on buildings and monuments. The adults fly to lights and enter dwellings through gaps.

Removing the breeding source is the best way to control midges. Management is site-specific; pesticides are generally not useful. Manipulating lights to shine away from buildings will reduce midge attraction.

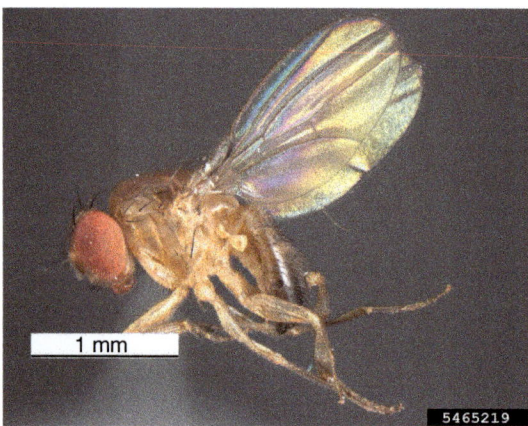

Fig. 3.15. Typical features of an adult fruit fly. (Courtesy of Pest and Diseases Image Library, Bugwood.org.)

Fig. 3.16. Typical features of an adult midge. (Courtesy of David Cappaert, Bugwood.org.)

4 Household Pests and Their Control – Mosquitoes

4.1 Mosquitoes in General

A number of diseases are linked to mosquitoes and mosquito bites (Table 4.1). There are over 3000 species of mosquitoes, of which only a few feed on human blood. Control measures against mosquitoes are generally directed against these few species.

● *Mosquitoes remain distinct from other insects owing to their sucking mouth parts, blood sucking habit and females laying eggs in water.*

Only female mosquitoes need a blood meal. Mosquito species show a clear preference for certain animals, which could be reptiles, birds or mammals. Humans have become a preferred host for a few species simply as an adaption and the ease of finding a blood meal. They are known to use body smell, carbon dioxide emitted from breath and emitted heat from the skin of the animal or human to locate the source. Some species prefer biting at certain hours, for example, at dusk and dawn or in the middle of the night; others may bite throughout the day.

Among the mosquitoes there are two major groups that suck human blood, which are described as follows (Table 4.2).

4.1.1 Life cycle

Adults

The mosquito life cycle is generally completed within 2 weeks in tropical conditions. Mosquitoes breed in water and go through four stages in their life cycle, which are: egg, larva, pupa and adult (Fig. 4.1). The females usually mate only once and produce fertilized eggs throughout life at intervals. In order to produce eggs, female mosquitoes require a blood meal. The blood meal helps egg development, which can take up to 3 days. The time taken may vary based on the temperature. Once eggs are developed the females look for a water body in which to deposit their eggs. Egg laying is followed by subsequent blood meals and another egg laying. Mosquito males, however, do not feed on blood and live on plant juices.

Eggs

Mosquito females lay a variable numbers of eggs over their life span. The number varies between species, and could be around 100 per laying. The nature of egg laying also varies among species as some lay their eggs directly on the water surface either singly (such as *Anopheles*) or stacked together in groups called rafts (such as *Culex*). *Aedes* species mostly lay their eggs just above the water line on artificial or natural water-containing items. These eggs hatch only when flooded with water. Also, uniquely, these eggs can remain viable for a number of weeks, even months, if left dry.

Larvae

Eggs hatch into larvae. These larvae go through three or four different stages before becoming pupae. The larvae have distinct characteristics that can be used to identify them (Fig. 4.6). Species show distinct variation in the size of their breathing apparatus, the siphon located in the tip of their abdomen. Both *Aedes* and *Culex* have well-developed siphons, but *Anopheles* larvae have a rudimentary siphon. Larvae move around between the surface and the bottom to breathe and feed on plankton and small organisms. Eventually, larvae develop into pupae.

Pupae

Larvae develop into comma-shaped pupae. The pupa lives in water and, like the larva, is quite active. The pupal head and thorax are greatly enlarged and a pair of respiratory tubes project from the body cover. The pupal stage generally lasts 3–7 days depending on climate. Pupae spend most of their time clustered together on the water surface.

Table 4.1. History of human diseases linked to mosquitoes.

Year	Discoveries of mosquito-vectored diseases
1897	Malarial parasites seen to develop in mosquitoes
1899	*Anopheles* sp. are the vector of human malaria
1900	Transmission of yellow fever by mosquito *Aedes aegypti*
1902	Transmission of dengue fever by mosquito

Table 4.2. Two major mosquito groups.

The Anopheline group		The Culicines group	
Anopheles	*Culex*	*Aedes*	*Mansonia*
The genus Anopheles is recognized for its role in transmitting malaria	Vectors of filariasis and some viral diseases	Vectors of dengue fever, yellow fever, chikungunia, zika and other viral diseases	Vectors of brugian filariasis

Fig. 4.1. Typical features of an adult mosquito laying eggs on a water surface. (Courtesy of Susan Ellis, Bugwood.org.)

They are a non-feeding stage. Once fully developed, the pupa transforms into an adult mosquito.

4.2 *Aedes* Mosquito

There are over 950 species *of Aedes* mosquitoes around the world (Rozendaal, 1997). Among them a few species of *Aedes* are now considered serious public pests (Fig. 4.2). Two species, namely, *Aedes aegypti* from Africa and *Aedes albopictus* from Asia have transported themselves across the world and are associated with viral diseases such as yellow fever, dengue fever, chikungunia, zika and many more.

4.2.1 Life cycle

Aedes eggs are laid singly on moist surfaces just above, near or on the water line in temporary pools,

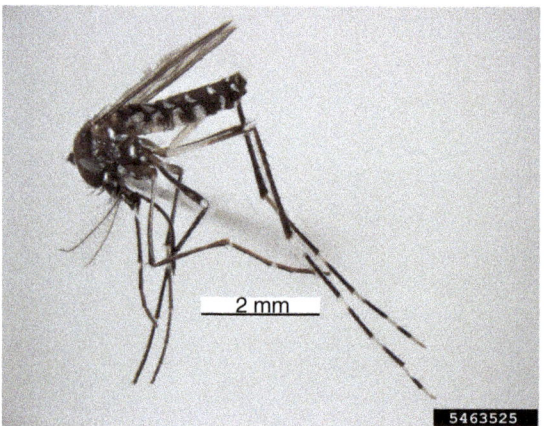

Fig. 4.2. Typical features of an adult *Aedes* mosquito, *Aedes albopictus*. (Courtesy of Pest and Diseases Image Library, Bugwood.org.)

containers and other clear water habitats. These eggs are unique in a way as they can withstand drying for many months and hatch only when flooded. Eggs are also known to survive in cold temperatures in regions with cold winters. Some species are also known to breed in coastal salt marshes and swamps that are flooded at intervals by unusually high tides or heavy rain, while other are adapted to breed in agricultural irrigation tanks (Rozendaal, 1997).

Aedes aegypti

- Identification: A bright silvery lyre-shaped dorsal pattern and white banded legs.
- Habitat: Bites, rests and lays eggs both indoors and outdoors.

Fig. 4.3. (a) *Aedes aegypti*. (Courtesy of Pest and Diseases Image Library, Bugwood.org.)

Fig. 4.3. (b) *Aedes albopictus*. (Courtesy of Pest and Diseases Image Library, Bugwood.org.)

- Breeding: The major breeding places are human-made containers.
- Biting: Sneaky biter.
- Preference: Prefers blood meals from humans and, to lesser extent, from domestic mammals, which makes them a less capable vector of zoonotic diseases but more capable of transmitting human viruses.

Aedes albopictus

- Identification: A single longitudinal silvery dorsal stripe and white banded legs.
- Habitat: Mostly an outdoor (garden) mosquito.
- Breeding: Shows preference for tree holes and bamboo internodes with water but can also utilize human-made containers for its immature development.
- Biting: Aggressive biter.
- Preference: Bites humans and a variety of available domestic and wild vertebrates, which makes them a more capable vector of zoonotic diseases and less capable of transmitting human viruses.

Behaviour

The biting periods for *Aedes* mosquitoes are mainly during dawn and dusk hours. However, the indoor environment with constant lighting and temperature in addition to the presence of humans throughout enables mosquitoes to be active most of the time and bite anytime.

4.3   *Culex* Mosquito

There are around 550 species of *Culex* around the world, most of them from tropical and subtropical regions. *Culex* is mostly a nuisance pest, but some species are important vectors of diseases like filariasis and Japanese encephalitis (Rozendaal, 1997).

4.3.1   Life cycle

Culex females lay eggs in clusters called rafts. Each raft may contain 100 or more eggs (Rozendaal, 1997). Eggs are laid on water surfaces. The rafts float until they hatch in 2–3 days. *Culex* species breed in a large variety of stagnant water pools, from artificial containers to natural water bodies, but especially drainage systems in urban areas. The common species *Culex quinquefasciatus* is a major nuisance and vector of filariasis, and breeds especially in polluted water bodies with organic material, such as septic tanks and open drains (Fig. 4.4). *Culex quinquefasciatus* is the commonest species found in rapidly expanding urban areas in developing countries, where drainage and sanitation are inadequate (Rozendaal, 1997).

Behaviour

Culex quinquefasciatus is an urban and domestic species. The adult females bite people and animals throughout the night. They are active indoors and outdoors. During the day they are inactive and are often found resting in dark corners of rooms, stores, shelters and shrubs, culverts, edges

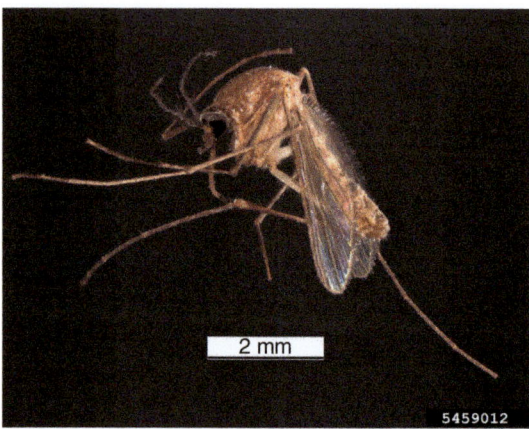

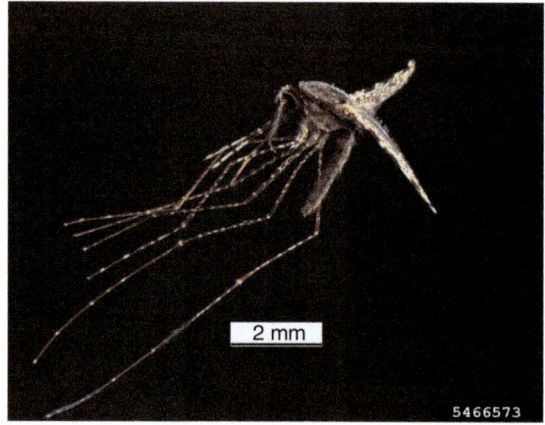

Fig. 4.4. Typical features of an adult *Culex* – *Culex quinquefasciatus*. (Courtesy of Pest and Diseases Image Library, Bugwood.org.)

Fig. 4.5. Typical features of an adult *Anopheles* – *Anopheles annulipes*. (Courtesy of Pest and Diseases Image Library, Bugwood.org.)

of drains and any places with shade and darkness (Rozendaal, 1997).

4.4 *Anopheles* Mosquito

Over 300 species of *Anopheles* have been recorded around the world, of which 60 of them are attracted to humans (Rozendaal, 1997).

Appearance

The most useful characteristics for distinguishing *Anopheles* species (Fig. 4.5) from other mosquitoes are the following:

- The length of palps and proboscis are the same.
- During the resting state, their mouthparts and abdomen are in a straight line and at near right angles to the resting surface.

4.4.1 Life cycle

Anopheles lay eggs singly in clean water pools, mostly natural with or without vegetation, or algae. *Anopheles stephensi* in South Asia is an exception as it is adapted to lay eggs in containers such as pots, tubs, cisterns and overhead tanks with clean water. Hatching occurs in a few days, the larvae float in the horizontal position on the surface, where they feed on small organic particles. In tropical conditions the duration of development from egg to adult is 2 weeks (Rozendaal, 1997).

Behaviour

Anopheles mosquitoes are active at night. Each species has a specific peak biting hour, and there are also variations in their preference for biting either indoors or outdoors. Different species also show preferences between human and animal blood. The *Anopheles* species flying in from resting places to feed on humans usually enters houses, takes a bite and rests indoors for a few hours before flying out. In some cases they may spend their entire time indoors digesting their blood meal and producing eggs. Once the eggs are fully developed the gravid mosquitoes leave their resting place and try to find a suitable water body in which to lay eggs (Rozendaal, 1997).

4.5 *Mansonia* Mosquito

Mansonia mosquitoes are commonly found around marshlands in tropical countries. Some species are important as vectors of brugian filariasis, which is restricted to southern India and parts of Malaysia (Rozendaal, 1997).

4.5.1 Life cycle

The females of this species lay their eggs in masses that are glued to plant parts hanging or floating over the surface of water. The larvae and pupae attach themselves to aquatic plants for the purpose of breathing. This species occurs only in water bodies containing permanent vegetation, such as

Mosquitoes: Characteristics Of Anophelines And Culicines

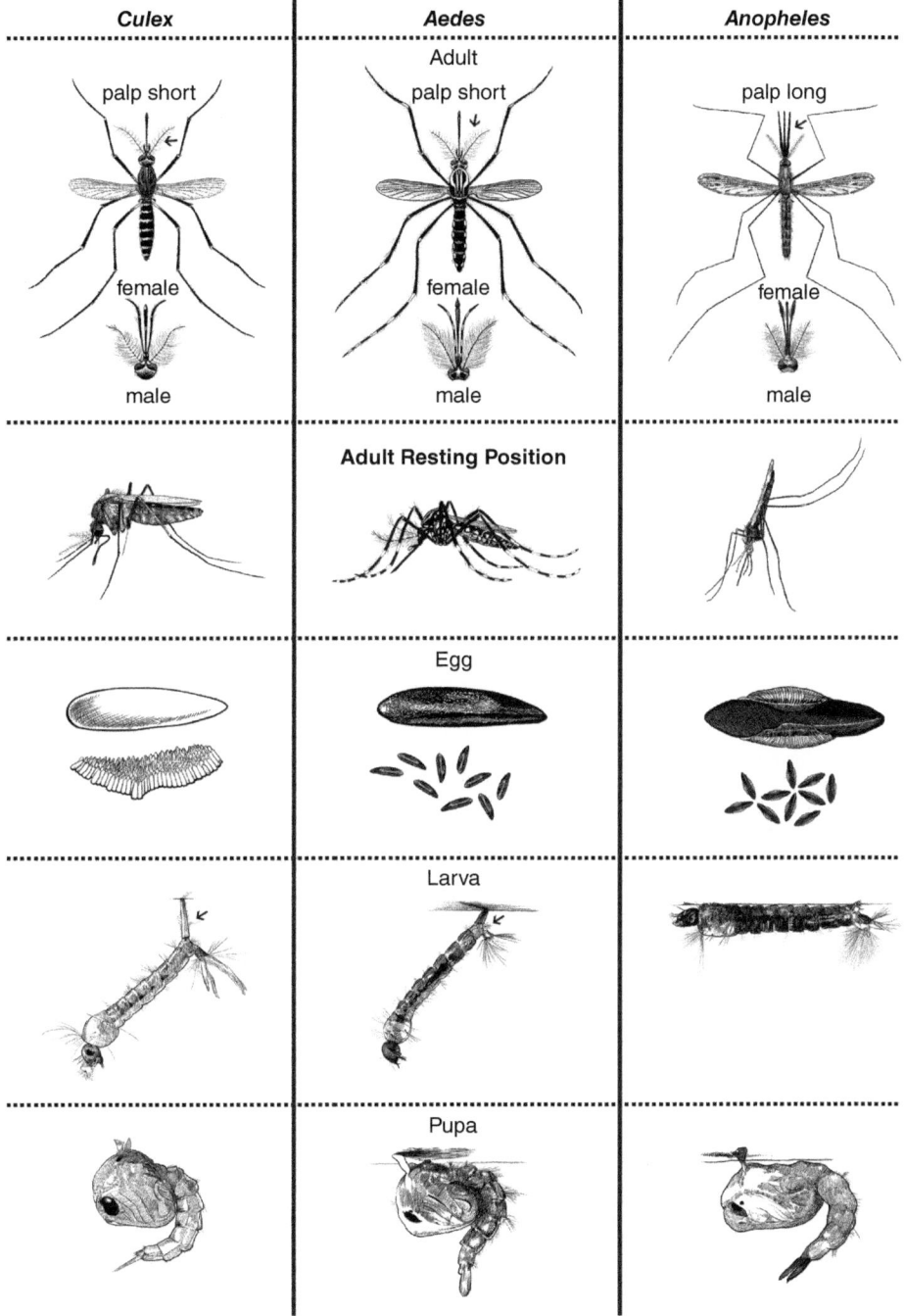

Fig. 4.6. Key identifying characteristics of mosquito species. (Adapted from *Pictorial keys to some Arthropods and Mammals of Public Health Importance*, US Department of Health Education and Welfare, Public Health Services, Washington DC, 1964.)

swamps, ponds, grassy ditches and irrigation canals (Rozendaal, 1997).

Behaviour

Mansonia species usually bite at night, mostly in outdoor locations, but some species enter houses also. After a blood meal females generally rest outdoors before laying their eggs.

4.6 Mosquito Habitats in Urban Areas

Mosquito breeding sites in urban areas are divided into three main types, namely:

- *Indoor breeding site with clean water*: These include water storage for washing and drinking, floor vases, uncovered drains, water pipes with holes, etc. preferred by *Aedes* species.

- *Outdoor breeding site with clean water*: These are mainly in-filled receptacles, such as containers, tree holes, and roof gutters, tyres, pots, ditches, etc., which serve as breeding sites for *Aedes* species. *Anopheles stephensi*, a vector of malaria in some urban areas in South Asia, often breeds in wells, ponds, cisterns and containers used for the storage of water.

- *Outdoor breeding sites with polluted water*: These are mainly sites such as drains, canals, exposed septic tanks, manholes, etc., which serve as favoured breeding sites for *Culex* species (Fig. 4.7).

4.7 Control and Management of Urban Mosquitoes

Mosquito control is challenging work for the very reason that it involves a number of stakeholders (participants) including homeowners, building or

Fig. 4.7. Public breeding grounds for *Culex*. (Courtesy of Biosav.)

area managers, municipalities and, at times, the local government. The work is also continuous and needs thorough monitoring.

On a building or home level, which is very different from municipality or area-wide mosquito control work, the following methods are recommended:

- Improving sanitation and water management by removing water-accumulating objects indoors as well as in the neighbourhood, such as by clearing roof gutters, potted plants, drains and water-accumulating objects, etc.
- Improving construction and design to prevent water accumulation on roof tops, etc.
- Modifying the life styles of inhabitants such as by wearing appropriate clothing, sleeping under a bed net, covering water storage areas, etc.
- Installing exclusion devices to reduce host–vector contact, such as screens, shutters, etc.
- Regular larviciding of water-accumulating areas indoors and outdoors.
- Adulticiding areas with active or breeding populations indoors and outdoors.
- Residual treatment of resting places both indoors and outdoors.

4.7.1 Recommended insecticide application

Regular larviciding in water-accumulating areas

1. *Using larviciding granules* for treating all water bodies, both indoors and outdoors. If the larvae, such as those belonging to *Anopheles*, are the target pests, then a formulation with a surface spreading characteristic should be chosen. Other larvicide formulations generally drop to the bottom and release the active agent from below.

To treat structures, these granular larvicides can be added to objects like flower vases, roof gutters, tree holes, drains, canals, unused toilet bowls, floor drains and all obvious and suspected breeding sites outdoors not used or visited by wildlife. Precautions should be taken not to treat drinking water or water for human use such as that used for washing and bathing as well as water with aquatic life such as fish aquariums. For water bodies like canals and ponds, the water area along the edges of the water bodies is most suitable for application (Fig. 4.8). An example formulation is temephos.

Fig. 4.8. Public areas with stagnant water are breeding sites for mosquitoes. Such a situation can easily be treated using granular larvicides. (Author's photo.)

2. *Larviciding formulations* containing an insect growth regulator (IGR) can be applied using handheld or backpack sprayers on small, mid-sized to large still water bodies. Precautions should be taken to ensure that these bodies do not contain fish and other forms of aquatic life. The area along the edges of the water bodies 1–3 m from the shore line is most suitable for application (Fig. 4.9). Example formulations are pyriproxefen or diflubenzuron.

3. *Larviciding oil* can be applied over outdoor water bodies to physically suffocate breeding mosquito larvae and also to prevent adult females from laying eggs. Self-spreading larviciding oils are applied on the edges of water bodies and then spread over large sections. Its mode of action is physical, rather than chemical, and it works by lowering the water surface tension that affects all aquatic stages of the

Fig. 4.9. Using a liquid larvicide formulation to spray on stagnant water in a public area. (Courtesy of Biosav.)

mosquito life cycle. Example oils are kerosene or diesel oil.

4. *Monomolecular surface films (MMFs)* are non-ionic surfactants. These act in a similar way to all self-spreading mosquito oil formulations to form a uniform film as a monolayer. This product requires a dosage many times lower than petroleum-based oils. An example is silicon-based MMFs.

5. *Biological formulations* in the form of tablets, briquettes, powder or liquid can be used as larvicides. These products are highly specific to mosquito larvae only and will not harm any non-target organisms. For example, *Bacillus thuringienensis*, Spinosad.

Adulticiding in breeding and resting areas

1. *Insecticide treatment with a non-residual formulation* both in indoor and outdoor areas can achieve very quick control of adult mosquitoes. Thermal fogging machines, truck-mounted or backpack mist blowers or handheld ULV (ultra low-volume) applicators can be used depending on the location and area to be covered. This work requires job site preparation and prior intimation to the client. The best time to do this work is generally early morning or early evening if outdoors. Indoor treatment can be carried out at any suitable time (Figs 4.10 and 4.11). For example, pyrethroid-containing formulations.

2. *Insecticide residual treatment* on resting surfaces such as walls, surfaces of drains, under bridges, along canals, bushes and on garden plants can be done any time of the day. A suitable backpack sprayer or a ULV mist blower can be used to coat the surfaces thoroughly to the point of running. A residual insecticide formulation should be used (Fig. 4.12), for example, pyrethroid-containing formulations.

Fig. 4.10. Using handheld low-volume applicator to treat space in inaccessible areas. (Courtesy of Biosav.)

Fig. 4.11. Truck-mounted ULV machines are useful for achieving area-wide coverage in a short time. (Courtesy of Biosav.)

3. *Mosquito traps*: Adult mosquito traps using attractants like CO_2 gas, octenol and temperature have been useful in mass capturing flying mosquitoes in urban areas (Fig. 4.13). However, these devices face limitations as they cannot eradicate the mosquitoes totally and are often expensive to maintain.

Fig. 4.12. Insecticide residual spray (IRS) used to treat solid surfaces such as house walls to kill resting mosquitoes. (Courtesy of Biosav.)

Fig. 4.13. Outdoor mosquito traps are useful devices to minimize isolated mosquitoes. (Author's photo.)

5 Household Pests and Their Control – Bed Bugs

Bed bugs have emerged as an important pest in recent times, mostly in subtropical and temperate countries. They are commonly reported from both high-end and public low-income residence buildings, and such structures as hotels, resorts, hospitals, restaurants, stores and homes. There are two species of bed bugs that feed on humans. They are the common bed bug *Cimex lectularius*, which occurs in most parts of the world, and the tropical bed bug *Cimex hemipterus*, which occurs mainly in tropical countries in Asia and Asia-Pacific. They cause severe nuisance when they occur in both small and large densities. Unlike most blood feeders, bed bugs are not important in the transmission of any disease.

Appearance

Bed bugs have a very characteristic flat and oval-shaped body with no wings (Fig. 5.1). The average length can be between 4 and 7 mm. They are shiny reddish-brown in colour and sometimes may look dark brown. There are two common species of bed bugs, namely, *C. hemipterus* and *C. lectularius*, and the morphological differences between the two species are very subtle. The difference is a broader prothorax (a region behind the head) for *C. lectularius* compared to *C. hemipterus*. Also *C. lectularius* is shorter than *C. hemipterus* in general.

5.1 Bed Bug Life Cycle

There are three stages in the bed bug life cycle, which are egg, nymph and adult. Eggs are white and about 1.0 mm long. The nymphs, once hatched, look just like adults but smaller. Complete development from egg to adult takes from 6 weeks to several months depending on the temperature and the availability of food. There are five nymphal stages. Both males and females feed on the blood of unsuspecting humans, mostly when they are at rest. In the absence of humans they feed on mice, rats, chickens and other animals. Feeding takes about 5–15 min

for adults, less for nymphs, and is repeated about every 3 days. Adults can survive for a year without food.

Between meals they hide in dark dry places and in the vicinity of the host, such as box springs, parts of the bed frame, headboard of beds, mattresses, cracks on walls, floor cracks, furniture frames, upholstery, behind picture frames, electrical sockets, carpets and curtain poles. Usually hiding places are also used for breeding.

Dispersal

Bed bugs do not have wings and travel short distances by themselves. In a building with many suitable hiding places they crawl from one room to another. At times they are passively shifted by movement of furniture, furnishings and household items. They spread from one building to another mainly through second hand furniture, furnishings, bedding, clothes or any item closely associated with human use. In commercial buildings such as hotels and resorts bed bugs enter through luggage, luggage carriers and used goods in addition to the above-mentioned items.

5.2 Control and Management of Bed Bugs

Bed bugs can move rapidly when disturbed and need a thorough inspection before any control measure is undertaken. Poor implementation of control measures may cause disturbance, spread or relocation of the pest. Control measures should be carried out only if there is evidence of the presence of the pest.

Locating infestation

Detecting the presence of bed bugs is very important in a bed bug management programme. In fact, early detection of bed bugs is perhaps the single

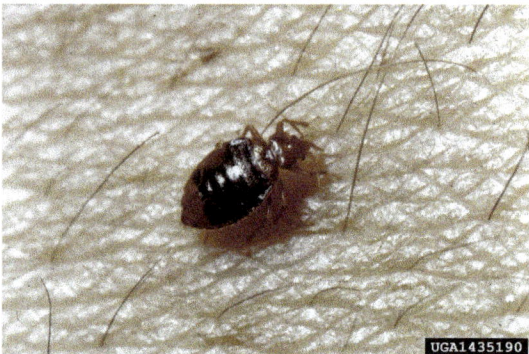

Fig. 5.1. An adult bed bug. (Courtesy of Clemson University – USDA Cooperative Extension Slide Series, Bugwood.org.)

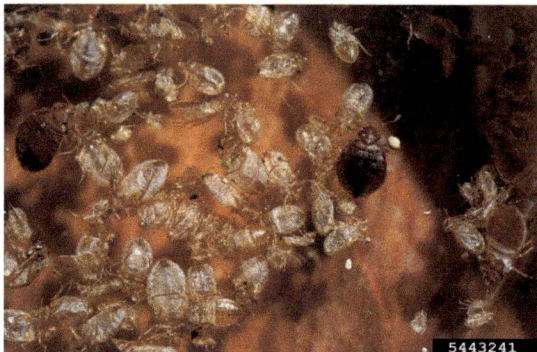

Fig. 5.2. Adult bed bug with cast skins in a highly infested location. (Courtesy of Whitney Cranshaw, Colorado State University, Bugwood.org.)

most important factor in eradicating bed bugs in an efficient and cost effective manner.

Methods for monitoring bed bugs fall into the following categories:

- Client interview and inspecting bite marks.
- Visual inspection in suspected locations.
- Placing monitors.
- Using aids such as trained detection dogs.

A combination of several methods generally provides better detection than relying on a single method.

Conducting an interview

The presence of bed bugs can also be determined by conducting an interview with the residents as well as examining bite marks if made available. In such situations it is best to look for the following:

- Bites in tight lines of multiple, small, red or pink marks with a central dark spot.
- Bites in the form of red welts (small, flat or raised).
- Bites with distinct swelling.
- The bites are usually on the upper body, from the waist up.
- However, it is best to document the bites and consult a physician or an expert to get a correct opinion.

Visual inspection

Infestation can be detected by the examination of possible hiding places for the following:

- presence of live bugs;
- cast-off nymph skins, eggs (Fig. 5.2);

- clusters of dark brown or black spots of dried excrement on infested surfaces (Fig. 5.3).

Bed bugs also exude a subtle, sweet, musty odour. Houses with large numbers of bed bugs may have a characteristic unpleasant smell. Live bugs can be detected by spraying an aerosol of pyrethrum into cracks and crevices, thus irritating them and driving them out of their hiding places. At times hot air from a hair drier can also be used to drive them from their hideout.

5.2.1 Use of traps and monitors

Passive monitors

Several passive monitors have been developed to help detect bed bugs (Fig. 5.4). These monitors provide shelter and do not use any attractant to work. They must be used directly under beds, furniture legs or any area where humans spend time resting. Despite the fact that the device is passive in itself, the presence of the human host provides this device with the most effective natural lure available.

Several designs of passive monitors consisting of layered or corrugated cardboard are commercially available. They offer alternative harbourages to bed bugs looking for a hiding place. The monitors are placed on or near beds, couches, daybeds, sitting and resting areas. Infestations are identified by the presence of bed bugs or their faeces and cast skins.

- *Often shiny casts of other insects such nymphal German cockroaches or booklice may be confused with those of bed bugs.*

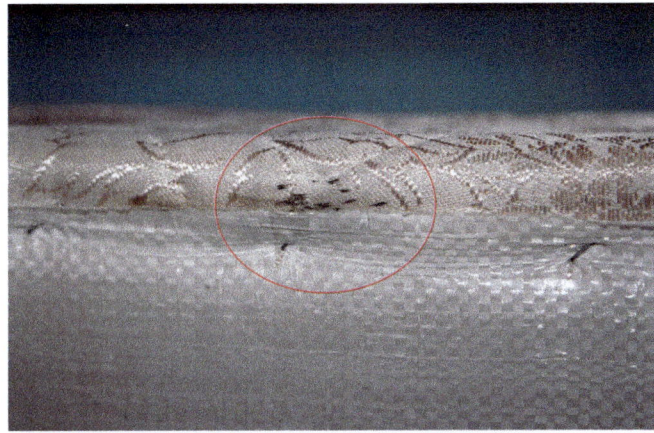

Fig. 5.3. Bed bug spotting mark on the side of a mattress indicating presence of active pest. (Courtesy of Dr Stephen Doggett.)

Fig. 5.4. Passive monitors provide excellent harbourage when used close to human resting places such as beds or couches. (Author's photo.)

Active monitors

Bed bugs can be attracted and trapped by baited pitfall traps. These monitors contain attractants and they can be used in both occupied and vacant rooms. Results can often be obtained overnight. Many field tests have shown carbon dioxide (CO_2) to be the most important lure when compared with heat and chemical attractants and it is believed to be the key attractant in an effective active monitor. However, a combination of multiple attractants can be more useful but may make the traps expensive.

5.2.2 Recommended insecticide treatment

Based on the nature and location of the infestation, the type of treatment can be chosen. The following are recommendations:

1. *Spot treatment* is recommended for light infestation once located. The treatment starts by separating the infested items from the rest. Infested items can then be packed into a plastic bag and sent for steam laundry, or a suitable heat treatment. Aerosol spray cans can be used to spray household insecticides onto mattresses, into crevices in walls and into other possible hiding places. Among the effective insecticides are the pyrethroids, propoxur, dichlorvos and neonicotinoids. Wettable powder or dust formulations have shown better results compared to emulsifiable concentrate (EC) or suspension concentrate (SC) formulations. Sealing the cracks and crevices with suitable plaster or sealant is recommended following the treatment.

Placing a dichlorvos strip in a large trash bag for 1–2 weeks with the infested items can kill all nymphal stages of bed bugs. Resin strips impregnated with dichlorvos (DDVP) are useful for treating small items, such as clothing, shoes, cushions, pillows, etc.

2. *Residual insecticide spray* is recommended for high as well as recurrent infestation. One treatment is normally sufficient to eliminate bed bugs but, if an infestation persists, retreatments should be carried out at intervals of not more than 2 weeks. In many countries, resistance of bed bugs to insecticides is becoming common. The insecticide selected should thus be one known to be effective against the target population. The addition of an irritant insecticide (e.g. 0.1% pyrethrin) helps to drive the bugs out of their hiding places. All cracks, crevices, furniture and bed frames need to be treated thoroughly after isolation of infested items such as mattresses, furnishings, carpet and clothing. Recommended insecticides for bed bugs should be used for such treatment with the correct dosage. Sealing of cracks and crevices should be undertaken to prevent unhatched eggs from hatching and coming out. A number of follow-up treatments are mandatory to evaluate the quality of the work.

3. *Dusts* such as pyrethroid-containing dust formulations have been found to be superior to spray formulations at times. Infested areas or items such as mattresses and box springs can be dusted carefully and wrapped in plastic. Pyrethroid dusts can be very irritating to the applicator, and require proper personal protective equipment during application. After application, the excess dust should be vacuumed to minimize human exposure to the dust.

5.2.3 Fumigation

Fumigating of all infested goods as well as suspected goods is recommended when the area has large and multiple location infestations. Goods are to be sealed in a container and fumigated using available fumigants in a fumigating chamber. Fumigation should be conducted only by trained and certified personnel in an unpopulated area. Fumigation is the only method that, when applied correctly, will result in 100% elimination of all bed bug stages from infested structures and/or infested items.

The fumigation process requires several days, which is often difficult and can be cost prohibitive for large structures (e.g. apartment buildings, hotels, etc.). Care must also be taken not to disperse bed bugs to new locations during the temporary relocation of the occupants and their items.

5.2.4 Heat treatment

Heat has long been used in bed bug control prior to the advent of modern insecticides. Boiling water, hot air and steam were used to treat mattresses, box springs, bed frames and other harbourage sites. The first two methods are not convenient or safe. Industrial room heaters can heat up rooms to near 60°C. A minimum of 52–54°C is sufficient to kill bed bug mobile stages and eggs with a 3–4 h exposure time.

A growing number of professionals, as well as property maintenance staff, are using heaters as a safe and affordable bed bug control method. Hot air can kill all stages of bed bugs. Users need to adjust the air flow rate so that the room heats up quickly leaving no air pockets. Heaters with larger attachments such as axillary fans usually do this work well. It is also mandatory that a number of temperature readers such as thermocouples and sensors are used at different locations and heights in the treatment area to accurately measure the temperature.

- *Although heat is highly effective, it is only as effective as its ability to penetrate.*

For homes with a large amount of furniture and other items, whole-house heat treatment or custom-made heat chambers are used to control bed bugs. Achieving the minimum threshold temperature (54°C) in all areas including piles of clothing and deep inside furniture is essential. One hundred per cent elimination is less likely in severely cluttered conditions (especially with densely packed clothing) or in structures where concrete construction may serve as a heat sink, making it difficult to achieve lethal temperatures throughout the entire structure.

Steps for conducting heat treatment:

- Seal the room to make it airtight.
- Move the items to the centre of the room.
- Set up thermocouples/heat sensors at different locations and heights.
- Start up multiple room heaters to heat the room quickly.
- Use fans to circulate the heat evenly.
- Once optimum temperature is reached, stir/move/invert the room items.
- Keep the room at the desired temperature for 4 h before evacuating.

5.2.5 Personal protection

Repellents available commercially can be used against bed bug bites. Diethyl toluamide (DEET) is a commonly used insect repellent that can last 6–8 h. Spraying shoes, trousers, luggage and other vulnerable surfaces with repellent containing at least 10% DEET may prove effective in reducing the risk of contracting bed bugs directly.

6 Household Pests and Their Control – Termites

Termites are cellulose (wood) feeding insects. They feed on all types of wood: processed wood, raw timber, products made of wood, paper, textiles, plant roots, litter, soil humus, etc. The termite is a tropical pest, and ranks on the top of the list of pests for practitioners in Asia, Asia-Pacific, parts of Africa, the USA and Australia.

Termites are generally categorized into the following types: subterranean, drywood and damp wood (Table 6.1).

6.1 Subterranean Termite

Subterranean termites are a major pest of structures. The pest enters a structure mainly through cracks in the foundation or floor slabs, through wall gaps or expansion joints. They also enter using utility pipes such as electrical and telephone conduits, water and drain pipes. These termites also find above-ground routes to enter a structure by forming mud tubes over walls, adjacent trees or fences.

- *Termites are social insects, living in a colony with a distinct caste system and they digest wood by processing it through use of symbiont or fungal garden.*

Subterranean termites live in permanent nests also called mounds or termitaria. Mostly these nests are buried under the ground, not visible on the surface. Sometimes these nests are constructed above ground, and on trees or abandoned structures. Termites are capable of travelling far from their nests in search of food. Termites seek a combination of moisture and carbon dioxide emitted from mouldy, decomposing wood to find food.

Among several species of termites known to be pests, the most common ones are the species belonging to the genera *Coptotermes*, *Macrotermes*, *Microcerotermes*, *Microtermes*, *Globitermes*, *Nasutitermes*, *Odontotermes* and *Reticulitermes*. The species composition differs depending on the region of the world.

6.1.1 Life cycle

Termites are social insects, living in a colony with a distinct caste system unlike other insects (Fig. 6.1). A new colony of termites always starts from the reproductive or alate castes. Alate castes consist of males and females that leave the nest/mound at a certain period of the year in large numbers. This process is called 'swarming'. The swarming alates are attracted to light in buildings and homes. This swarming flight is called 'nuptial flight': where they find mates and establish a new colony. These mating alates then drop to the ground and immediately shed their wings. The mated pairs then excavate into the soil, or moist timber, hollow gaps in buildings, etc. and build a nest. The female, now a queen, continues to lay eggs at a rapid rate and soon a stable colony is established.

The different castes in a termite colony are morphologically and functionally distinct. Eggs hatch into white-coloured nymphs, which are capable of developing into any caste depending upon the requirements of the colony. Worker termites are pale cream in colour; soldier termites are the same size and colour, however their heads are much enlarged (almost half their body length) with noticeable black jaws/mandibles. Workers construct the distinctive shelter tubes and collect food to feed the young and other members of the colony. Soldier termites are responsible for guarding the colony and its occupants.

In general winged reproductive termites do not appear until the colony is 3 or 4 years old.

6.1.2 Termite castes

Termites are social insects and live in nests that serve to protect the colony, store food and maintain an optimum environment for growth and development. A termite colony can have from a few thousand

Table 6.1. Types of termite.

Subterranean termite	Drywood termite	Damp wood termite
Termites living in soil and using mud tubes to move above ground. They mostly use cracks or construction gaps to enter a structure. Always characterized by presence of soil in the area of infestation	Termites living in moistureless wood with no connection with the soil. They form nests in the structure that they are infesting, such as door and window frames, furniture, etc. Characterized by presence of pelletized droppings in the infested area	Termites living in moist wood such as fence poles, dead trees, etc. They may or may not maintain a connection to the soil. Rarely infest indoor structures

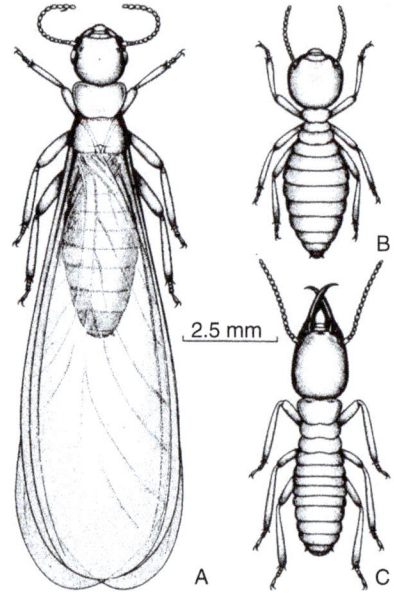

Fig. 6.1. Typical termite castes, (A) alate/reproductive; (B) worker; and (C) soldier.

Fig. 6.2. A typical colony consisting of queen, king, workers, soldiers, nymphs and eggs. Alates are found only seasonally. (Author's photo.)

to a few million individuals at a given time (Fig. 6.2). The functioning of the colony is governed by a system of self-regulated behaviour of individuals and division of labour among the different members of the castes. A colony harbours the following castes:

- Primary queen – usually one per nest, can be multiple – role is to produce eggs.
- Secondary queen/s – queen in waiting if the primary queen dies or is disabled.
- King – single or multiple per nest – mates with the primary queen(s).
- Soldiers – possess sharp mandibles, defend the nest from outsiders such as ants.
- Workers – travel to find food, feed nymphs, caregivers to eggs and other members.

- Nymphs – young termites that can develop into either soldiers or workers.
- Reproductive, alate or swarmer – sexually mature males or females that have wings to fly out of the nest and mate. Later become king and queen to start a new colony.

6.1.3 Common subterranean termite species as pests

These are generally classified into two groups (Table 6.2). A few notable termite species are described as follows:

Coptotermes

APPEARANCE. *Coptotermes* is a genus of termites in the family Rhinotermitidae. The genus *Coptotermes* is characterized by the presence of a pear-shaped head, narrow at the front with a pointed labrum in the soldier caste. Mandibles are slender, sharply

Table 6.2. Characteristics of lower and higher termite groups.

Lower termites	Higher termites
Family: Rhinotermitidae, Kalotermitidae, etc. Characteristics: Lower termites maintain a symbiotic relationship with gut protozoans, which help digest the cellulose material eaten by the termite For example, *Coptotermes* species	Family: Termitidae Characteristics: Higher termites do not have gut protozoans and instead depend on their own enzymes. This group is known to grow fungus on its faecal waste (combs). The termite eats the fruiting body of the fungus. They also eat the partly digested comb material For example, *Macrotermes* species

pointed and slightly incurved without marginal teeth. Most distinctive in the soldier caste is the large fontanelle (opening) at the front of the head, which exudes a white defence secretion when the insect is disturbed. *Coptotermes* have been shown to possess, as for other members of the Rhinotermitidae, sunken pores on their legs, which may produce a defensive secretion against predators. There are about 71 species, many of which are economically destructive pests. The genus is thought to have originated in Asia.

Four species under the *Coptotermes* genus are most common as pests (Figs 6.3–6.7). Morphologically they are very similar in description and can only be identified by experts by looking closely into the degree of mandible curvature and the width of the head at the base of the mandibles, and other features such as cuticular hydrocarbon and, recently, DNA.

BEHAVIOUR. In tropical conditions swarming can take place throughout the year under warm moist conditions. The colonies of the species can be large, from half a million to 2 million, and can travel over 100 m from the nest for foraging.

ECONOMIC IMPORTANCE. *Coptotermes gestroi* is the most common species in the genus as a pest in South-east Asia. The species originates in South-east Asia and is recognized from its origin, to territories such as Taiwan, Mauritius, Réunion Island, and across the Pacific Ocean to parts of Polynesia, Hawaii, Marquesas Island, Micronesia, Fiji, Mexico, Florida, the Caribbean Islands and is now spreading along the south Atlantic coast of Brazil. Reports of the species in Italy are also notable, although the infestation was reported in a yacht in Sicily. Similar reports of the species infesting sea vessels in the Philippines provide insights into how the species has moved across the world. Other notable pest species belonging to this genus are *C. formosanus*, *C. curvignathus* and *C. acinaciformis*.

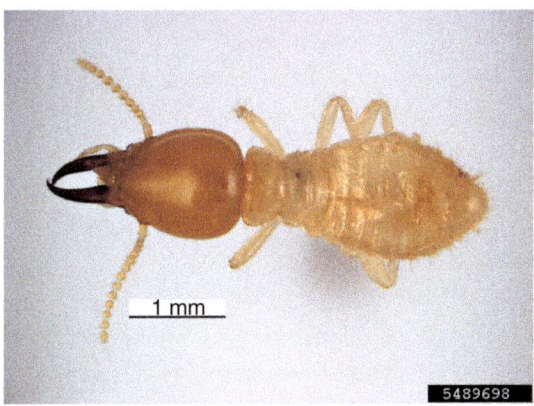

Fig. 6.3. *Coptotermes gestroi*, originated in South-east Asia and invaded the Pacific countries including the USA, Brazil and Mexico. (Courtesy of Pest and Diseases Image Library, Bugwood.org.)

Fig. 6.4. *Coptotermes acinaciformis*, possibly originated in Asia and moved to Australia. (Courtesy of Pest and Diseases Image Library, Bugwood.org.)

Fig. 6.5. *Coptotermes curvignathus*, originated in South-east Asia and invaded China. (Courtesy of Pest and Diseases Image Library, Bugwood.org.)

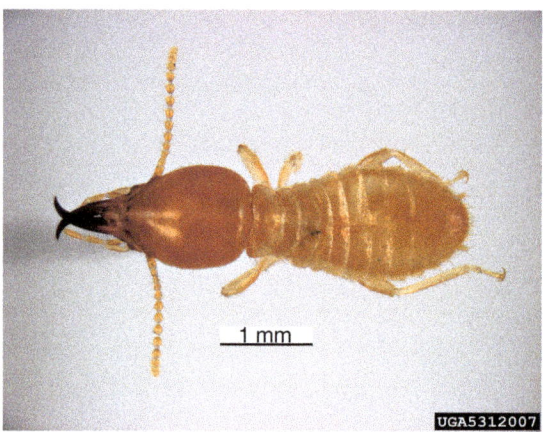

Fig. 6.6. *Coptotermes formosanus*, originated in China and Taiwan and invaded the USA, Fiji and Japan. (Courtesy of Pest and Diseases Image Library, Bugwood.org.)

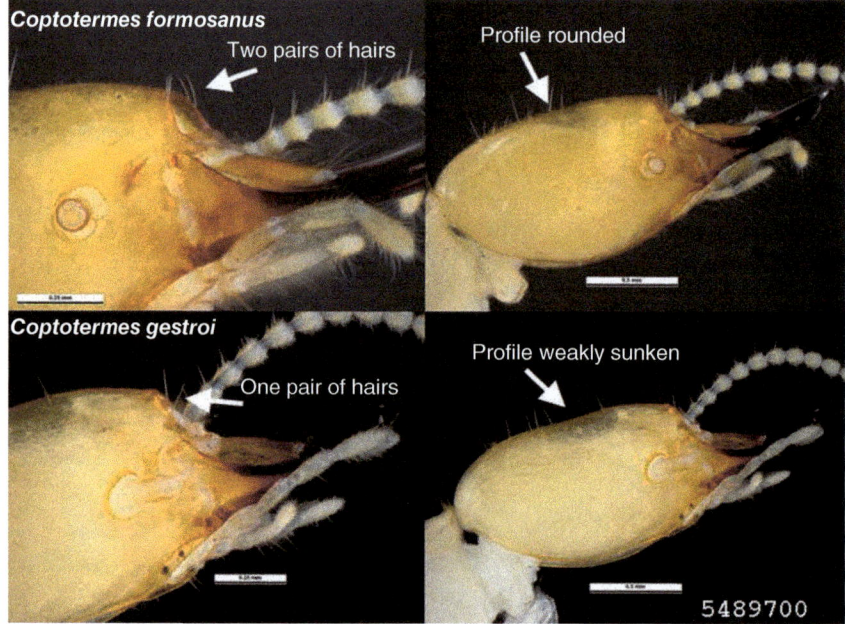

Fig. 6.7. Characteristic difference between *Coptotermes gestroi* and *C. formosanus* soldiers. (Courtesy of Pest and Diseases Image Library, Bugwood.org.)

In addition to its invasive nature, the species is known for its propensity to establish itself in urban areas, causing damage to wood and wooden structures used by man. The species also consumes all forms of wood commonly used in structures, including household items in addition to selective species of live trees. Reports of damage to fabrics as well as home fixtures and upholstery materials have also been noted. The species also shows a unique subterranean method of entering structures using subfloor cracks in concrete and brickwork. It is also known to make its way into a structure through expansion joints as well as electrical conduits, and regular service penetrations are common.

Damage resulting from a *C. gestroi* infestation can become severe in a relatively short time, especially when a structure is invaded by a large, mature colony. Dispersal flights, foraging tubes or damage are usually the first indications of an infestation. Advanced stages of infestation are indicated by the incorporation of nest material (carton) in hollowed wood or existing structural voids (Fig. 6.8) and aggregation (Fig. 6.9). All of these characteristics make *C. gestroi* a species of significant economic importance around the world.

Macrotermes

APPEARANCE. *Macrotermes* sp. is a common mound-building termite species found in the Asia-Pacific region (Fig. 6.10). The height of the mound varies from a few centimetres to over a metre. The species is a generalist feeder, feeding on various wood types, including barks, dried leaves and humus. Soldiers have a very distinct large head compared to their body. Both workers and soldiers show clear dimorphism, with distinct variation in body size – a distinguishing feature of the species (Fig. 6.11).

BEHAVIOUR. The genus is now a common feature of the suburban landscape in parts of South-east Asia, and is troubling pest control operators. With cities quickly growing into suburban areas, increased incidence of structural infestation and defacement of wood are coming to attention. Most of the observed entries to structures are through above-ground routes. Problems with moisture control are the dominant reason for the majority of the infestations.

The species belongs to the higher termite group. The termite is dependent on varied sources of plant material including wood. The chewed up plant matter and wood is used to construct elaborate fungus gardens in their nests, also known as 'combs'. The termites actually feed on both the fungus heads and the combs (Fig. 6.12).

The most intriguing feature of this termite, which remains unexplained, is the number of mounds the species makes in a given area. Often numerous active mounds can be observed on a small patch of land. Whether this species is capable of sharing food, nests and territories among adjacent colonies is yet to be investigated in this region. Another feature of the species commonly observed is the presence of multiple mature queens in a single chamber (Fig. 6.13). Each of these life history characteristics is a significant

Fig. 6.8. Characteristic cartoning, a sure sign of *Coptotermes* infestation. (Author's photo.)

Fig. 6.9. A typical highly infested *Coptotermes* site where both workers and soldiers are present in an above-ground situation. (Author's photo.)

reason for the success of this species as well as being reasons for concern in terms of pest management.

Microcerotermes

APPEARANCE. *Microcerotermes* sp. is a serious pest in urban areas of South-east Asia. The nests can be totally subterranean. Sometimes nests are built above the ground, on the trunks of living or dead trees, or on the top of stumps or fence posts and even inside buildings. The outside layer of the above-ground nests is like a carton: thin and easily penetrated or broken. The soldiers have clear rectangular-shaped heads – a distinguishing feature for species identification (Fig. 6.14).

Fig. 6.10. A typical *Macrotermes* mound in an open area. (Author's photo.)

Fig. 6.11. *Macrotermes gilvus* soldiers showing dimorphism with primary and secondary soldiers in one nest. (Courtesy of Josielyn Trinidad.)

The species is known to feed on a varied diet of tree bark, dead trees, leaf litter and, at times, wood and processed wood. Entry to structures is through above-ground routes or through gaps in construction (Fig. 6.15). The nature of damage can vary from severe to simple defacement of wood.

Nasutitermes

APPEARANCE. The soldier's head is drawn into a beak (Fig. 6.16). The species constructs arboreal nests in trees or on abandoned structures. The out-side layer of the above-ground nests is like a carton: thin and easily penetrated or broken. Subterranean mounds are also common at times in urban areas. The termites feed on decayed and weathered timber, particularly when it is in touch with soil. It is an occasional pest of fences, trees and weathered moist outdoor wooden structures (Fig. 6.17).

Globitermes

This genus is a common South-east Asian mound-building termite, nesting in the ground, predominantly

Fig. 6.12. A nest comb showing cultivated fungus, which serves as food for the worker termites. (Author's photo.)

Fig. 6.13. Multiple queens in one queen chamber. (Author's photo.)

Fig. 6.14. A specimen of *Macrocerotermes* sp. soldier with the typical rectangular-shaped head. (Courtesy of Dr P. Suresh.)

Fig. 6.15. *Microcerotermes* sp. above-ground nest. (Author's photo.)

in tropical rainforest. The most common species, *G. sulphureus*, has soldiers whose abdomens dehisce, covering any attacker with a sticky yellow-coloured defence chemical. Another distinct characteristic of this species is the presence of a bright yellow-coloured abdomen in the soldier caste, which has a salivary gland that extends to the end of the abdomen. This species is being noticed as an indoor pest of structures, infesting wooden structures in suburban areas of Malaysia.

6.2 Control and Management of Subterranean Termites

6.2.1 Termite damage

The presence of termites in buildings should be taken seriously. If the infestation is detected on a primary part of a structure such as a load-bearing column, roof trusses, staircases, floor or ceiling, immediate action is a must. A large colony of termites can damage wood very quickly, which can cause the structure to collapse and injure occupants.

Apart from primary parts of a structure, termites can damage all cellulose-containing household items.

They are also known to seriously deface wood and concrete surfaces, requiring expensive restoration work.

- *To prevent termites from causing damage, structure owners or homeowners should engage professionals to look for signs of an infestation, including termite droppings, discarded wings, hollow or damaged wood, mud tubes, pencil-sized dirt tunnels, etc.*

Prior to encountering one of these signs, or after a treatment, there are several preventive measures the practitioners should advise owners about that can help reduce the risk of an infestation. They are the following:

Fig. 6.16. *Nasutitermes* sp. soldier with characteristic head snout. (Courtesy of Pest and Diseases Image Library, Bugwood.org.)

- Minimize or monitor planter boxes around the structure.
- Trim all shrubs near home exteriors to allow airflow and quickly dry damp areas.
- Use products such as synthetic mulch or pea gravel when landscaping.
- Maintain home exteriors to prevent water from leaking into wooden siding and windows.
- Ensure crawl spaces are properly ventilated to minimize the amount of moisture around floor joints and subflooring.
- Ensure uninfested wood is used in any type of construction.
- Get the structure inspected annually by an expert for termite activity.

Termite entry into a structure

Termites mostly enter structures using concealed entry points (such as cracks and gaps in the structure). They may also use above-ground points, but this is less likely. In countries with termites as common structural pests it is important that adequate intervention methods are used when undertaking construction. Even though a host of parameters might encourage termites in the vicinity of the structure, a house constructed with no or minimal points of entry can remain uninfested for long periods. Foolproofing structures by using sound engineering methods is thus the best method for preventing termite entry. Recently the Asian market has seen the launch of a number of physical barriers, such as marine grade aluminium, stainless steel mesh and special grades of cement and resin mixtures in termite

Fig. 6.17. Mud tube used by *Nasutitermes* sp. to access wood in the structure. (Author's photo.)

proofing buildings. These products take care of construction gaps such as on the edges of slabs on the ground, cracks, and gaps in and around service penetrations when skilfully installed and between footings, floor and wall.

Predominantly, Asian houses are designed and constructed with steel and concrete. Wood is rarely used as the primary load-bearing item in a structure, but it is extensively used in secondary components such as ceilings, walls, partitions, wall cabinets and flooring. Walls are generally constructed with bricks or hollow blocks. Floors are constructed on sectional slab. Rarely, a single slab on ground (monolithic) method is used (Fig. 6.18). This is the primary reason for gaps and cracks in the floor, which make termite entry easier. The house wall in this construction type rests on the footings rather than on the floor. However, most commercial construction makes use of a monolithic slab and the walls are constructed directly on the floor slab (Fig. 6.19).

Termites gain entry into structures predominantly by using concealed points as in the following:

- Gap/s between footings and floor in non-commercial constructions (Fig. 6.20).
- Gap/s between floor slab and wall in commercial constructions (Fig. 6.21).
- Directly through cracks on the floor slab.
- Through service penetrations such as those made for water pipes and electrical wires. Above-ground penetration is also common.

Concrete slabs used as flooring have been shown to be an effective barrier to termite penetration.

However, to make it work as a complete barrier system, all penetrations and joints through the slab and the slab edges must be sealed, as these are the remaining locations that could allow concealed entry of termites. All service penetrations through a slab on the ground must be provided with a suitable termite barrier such as via the use of metal collars, mesh or cement mixtures. This is because

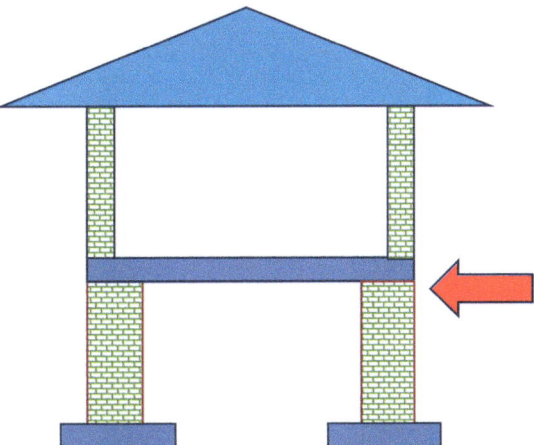

Fig. 6.19 Illustration showing commercial structures such as office buildings.

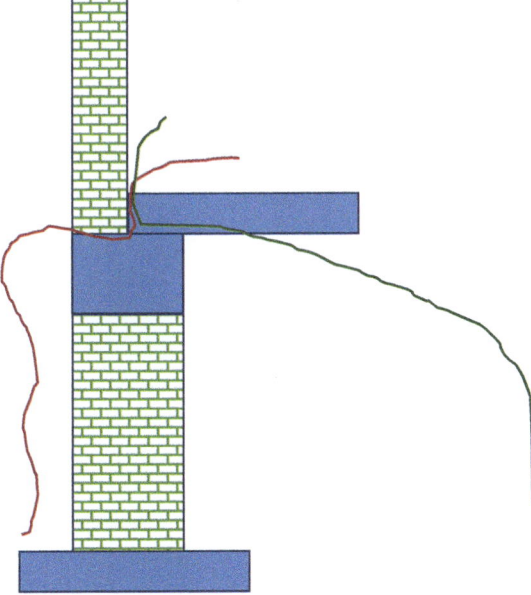

Fig. 6.20. Illustration showing possible routes of termite entry in non-commercial construction with sectional flooring.

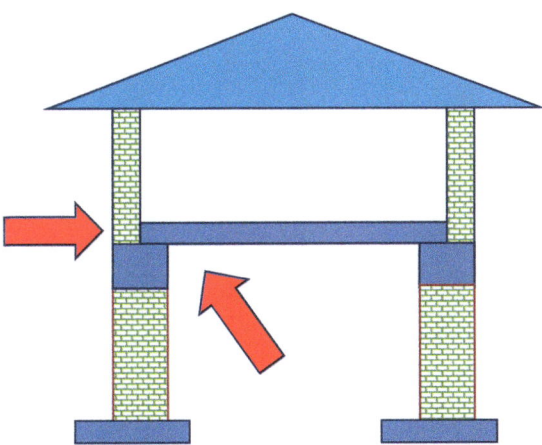

Fig. 6.18. Illustration showing non-commercial structures such as homes.

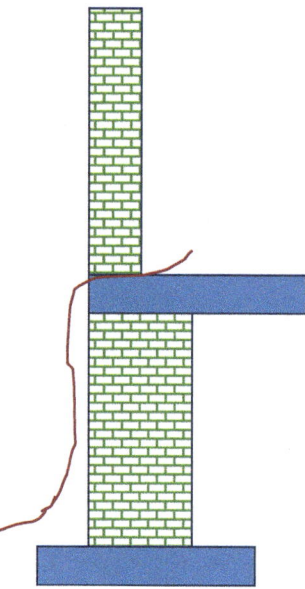

Fig. 6.21. Illustration showing possible route of termite entry in commercial construction with single slab flooring.

the gap that can open up between a pipe (or other penetration) and the concrete slab may be sufficient to allow termite entry. Studies have shown that a 1.4 mm crack is enough for termite entry.

What are the signs of termites in a structure?

- Swarms of winged reproductives flying from the soil or wood.
- Presence of wings discarded by termites.
- Damaged or discoloured wood.
- Hollow sound on tapping structural timber.
- Presence of soil and mud tubes.
- Presence of pellets or droppings near wooden items.

What types of structures are vulnerable to termite infestation?

A survey by the author has conservatively estimated 30% of structures in any Asian city to either have active termites or have had a previous infestation. Termite damage to structures is extensive due to poor preparation of builders against termites, absence of building codes specific for termites and the indifference of owners to safe practices. As a consequence of this most structures are infested, needing repeated treatments.

In Asia structures and buildings are designed and constructed with steel and concrete to resist drastic weather changes as well as typhoon-type winds and frequent earthquakes. Houses are constructed using slab-on-the-ground flooring. All commercial construction makes use of a monolithic slab, while houses often lay slabs that are laid in parts. Termites gain entry to the structure through cracks, joints, utility openings in the subfloor and sometimes by forming over-ground mud tubes. Wood is extensively used, most often unscientifically without considering termites as a major problem. Wooden flooring, baseboards, false walls, wall decorations, ceilings, staircases, doors and windows are common features of a structure. These are also the site where infestations start.

Prevalence of termites in a region makes all structures in the area prone to infestation. It is thus essential that some type of intervention method is used to prevent their entry into structures. However, in practice, very few structures are built with termite protection in their design. To overcome this, many countries have intervened and enforced building codes as mandatory. Adherence to such federal regulation has considerably improved structural protection. But most methods of intervention are challenged by nature and human activities, which continue to make structures susceptible to termites in their lifetime.

It often intrigues termite managers as to what could possibly make a structure favourable to infestation in a termite-prone area. Research has shown that a host of intrinsic factors govern termite foraging, which could determine final food selection. Predominant among them could be those factors that increase the overall fitness of the colony, such as distance from the nest and nutritional quality. It is also evident that termites use specific search methods to locate food and may use cues such as carbon dioxide and fungal emanations.

A retrospective analysis by the author (Dhang, 2011) showed that structures associated with the following have significantly more infestation:

- landscape;
- middle class owner; and
- constructed by developers.

Landscape and gardens around structures could serve as a source of moisture and food, attracting termites to the area, indirectly making the structure prone to attack (Fig. 6.22). Interestingly, structures constructed by developers showed significantly higher infestation than self-made structures (Fig. 6.23). This could be because developers often use unutilized lands with

Fig. 6.22. Landscape and planter boxes in close proximity to a structure are significant reasons for termite infestations. (Author's photo.)

Fig. 6.23. Houses built in newly developed areas by land developers are highly susceptible to termite attack. (Author's photo.)

natural undergrowth or agricultural land for building these structures. These areas are usually colonized by termites and eventually are an easy source of infestation. Most of these structures are part of the urban sprawl where the population has been relocating.

It is interesting to note that when all the three parameters key to infestation, namely, socio-economic, construction and landscape factors are analysed together, the significance of each parameter is evident.

- *In middle class structures, the construction method is a more dominant determinant to infestation than the presence of landscape.*
- *In contrast, among upper class structures, landscape is a more dominant risk factor than construction method.*

In spite of the limitations of having a smaller control group for comparison, the study elucidates the combined effect of key parameters in determining infestation. Presuming middle class structures have limitations in procuring superior quality materials, foolproof design and quality of construction methods become dominant factors compared to landscape. In upper class structures that have access to quality products and services, landscape becomes the determinant factor for infestation.

6.2.2 Termite control methods

Controlling termites begins with classification of the job. A termite control treatment job is classified as shown in Table 6.3.

6.2.2.1 *Post-construction – above-ground infestation with active termites*

This is the commonest job a practitioner will encounter. Clients and customers often call practitioners once they notice damage. A job with an above-ground infestation with active termites begins with identification of the termite species. Based on the species, the control strategy and method may vary. The following are descriptions of some common methods.

Control of above-ground infestation on a structure

Once an infestation and the number of infested spots are detected in a structure, the following methods can be used.

BAITING. Termite baiting is the best way to treat above-ground infestation (Figs 6.24 and 6.25). Baiting using a number of insecticidal active ingredients, such as bistrifluron, chlorfluazuron, hexaflumuron, noviluron, etc., has proven to be successful in controlling *Coptotermes* species. Chlorfluazuron-based bait in particular has shown evidence of controlling and eliminating other species of termites, including *Macrotermes*, *Microcerotermes*, *Odontotermes* and *Nasutitermes* species prevalent in Asia and Asia-Pacific. Similarly, bistrifluron-containing bait has been shown to control *Globitermes* species.

PROCEDURE. The detailed procedure is discussed on p. 100.

CHEMICAL TREATMENT. Treating the active spot with an insecticide spray is not a long-term solution for controlling the termite activity in the structure of concern. Chemicals such as organophosphates, carbamates, pyrethroids, neonicotinoids and phenylpyrazoles will kill active termites on the spot and prevent them from re-infesting the same spot for some duration until the residual effect of the chemical weathers away. However, the chances of the same termite colony resurfacing elsewhere in the structure/property are high. In this situation it is advisable that the entire perimeter of the house be treated using a soil treatment procedure in addition to the spot treatment. All forms of wood used in the structure should also be put under inspection and treated suitably. This method deters the colony from the structure, but does not kill the colony.

PROCEDURE. Treating the soil (commonly called soil treatment), building foundations and subslab areas as an industry practised intervention method is common to prevent infestation. The perimeter

Table 6.3. Classification of termite treatment.

Post-construction	Post-construction	Pre-construction
Above-ground infestation with active termites	Preventive on an existing structure without termite infestation	Protecting a future structure under construction

Fig. 6.24. Installing an above-ground baiting station on an active site. (Author's photo.)

Fig. 6.25. Above-ground bait station on an infested spot. (Author's photo.)

of a structure is generally treated by soil trenching or the rodding application technique (Figs 6.26 and 6.27). It is recommended that the rod spacing should be between 150 mm and 300 mm depending on soil type. Apply the chemical solution with suitable application equipment to form a complete and continuous chemical barrier (both vertical and horizontal) below the level of the floor or under the slab. The area of application may vary based on the nature of construction.

Concrete structures or structures with concrete slabs, aprons, tiles and driveways will need to be drilled to gain access to the soil. It is recommended that the drill hole spacing should be between 150 mm and 300 mm. Holes on both sides of the wall are highly recommended in cases where this is possible. However, at times holes inside the structure may

not be recommended. Once the chemical is injected through the drilled holes it is mandatory that the holes are sealed with cement.

Controlling above-ground mud tubes/nest

Some species make visible above-ground earthen mounds on structures. The most common ones are made by *Macrotermes* and *Microcerotermes*, common pest species in South-east Asia. Such mounds in the vicinity of a structure should be treated to prevent infestation. This author has found that physical demolition of mounds and treating the site with a residual chemical is a good method of control.

Above-ground mud tubes and mud trails on structures are also common. As termites are active in the mud tubes the baiting method is recommended. An above-ground bait station can be placed over the mud tubes and baited suitably (Figs 6.28 and 6.29). However, the success of the colony elimination is dependent on the intensity of baiting or frequency of bait replenishment. Small- to medium-sized colonies are easy to eliminate and can be totally controlled in 3 months from the time baiting starts. Larger colonies will require more time.

6.2.2.2 *Post-construction – prevention on an existing structure*

Preventive treatment of any structure can be undertaken either by the soil treatment method or by use of baiting.

Fig. 6.26. Treating the soil outside the structure using a power machine. (Courtesy of Andrew Gok.)

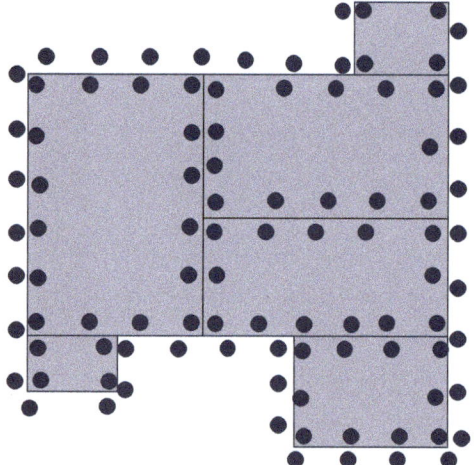

Fig. 6.27. Illustration showing the best way to perform a soil treatment is to inject the chemical on both sides of the wall of the structure.

Fig. 6.28. Positioning an above-ground station over a mud tube correctly before baiting. (Author's photo.)

Preventive treatment using the soil poisoning method

PROCEDURE.   Treating the soil, building foundations and subslab areas as an industry practised intervention method is common to prevent infestation. The perimeter of a structure is generally treated by soil trenching or the rodding application technique. It is recommended that the rod spacing should be between 150 mm and 300 mm depending on soil type. Apply the chemical solution with suitable application equipment to form a complete and continuous chemical barrier (both vertical and horizontal) below the level of the floor or under the slab. The area of application may vary based on the nature of construction.

Concrete structures or structures with concrete slabs, aprons, tiles and driveways will need to be drilled to gain access to the soil. It is recommended that the drill hole spacing should be between 150 mm and 300 mm. Once the chemical is injected through the drilled holes it is mandatory that the holes are sealed.

TRENCHING. This method involves digging a complete trench around the structure. The trench could be 20–30 cm deep and 20 cm wide. The solution of the termiticide in water is then allowed into the trench. Once the mixture is completely absorbed by the soil, the trench is filled with the soil and the same chemical solution applied to saturation point. The limitation of the method is that soil should be available all around the structure to form a complete trench.

Fig. 6.29. Adding bait to the station. (Author's photo.)

Preventive treatment using the baiting method

This method is used as a preventive method for termite treatment. In-ground stations (IGS) are placed in the ground around the exterior of the structure at intervals of 3–5 m. The distance between stations and the structure should be ideally as close as possible (1.0 m). However stations placed nearer or further away will also work. If soil is unavailable it may also be necessary to form holes in slabs or asphalt and place in-concrete stations. The structures treated with this method remain under protective monitoring. Once termites are detected in the station(s), bait is added and the colony is subsequently eliminated.

PROCEDURE. The detailed procedure is discussed on p. 100.

6.2.2.3 Pre-construction – on a structure under construction

Pre-construction termite treatment is done via the soil treatment method (Fig. 6.30). A chemical barrier is formed around various parts of the structure under construction such as the foundation wall, columns, etc. Soil around these parts of the structure is treated.

PROCEDURE. The standard treatment recommends that the subfloor area should be kept free of all debris such as trees, stumps, wood debris, mulch,

Fig. 6.30. Treating the soil at the foundation level before the start of construction. (Courtesy of Richard Tang.)

weeds, etc. Uniform distribution of the treating solution should be ensured so that chemicals toxic to subterranean termites can be used effectively to check termite infestation.

Treating the soil beneath the building and around the foundations with soil insecticide is a preventive measure, the purpose of which is to create a chemical barrier between the ground and possible cracks in the structure, woodwork or cellulosic materials in the building.

The barrier formed should be complete and continuous under the whole structure to be protected and should be in close contact with the foundations. Every part of the area treated should receive the prescribed dosage of chemicals. The standard states the treatment process is started when foundation trenches and pits are ready for mass concreting. Concreting should start when the chemical solution has been absorbed by the soil and the surface is dry. Treatment should be avoided on rainy days or when the soil is wet with subsoil water.

The treatment should also be applied to filled earth surfaces within the plinth area before laying the subgrade for the floor. Soil barriers, once formed, should not be disturbed. If they are disturbed, immediate steps should be taken to restore the continuity and completeness of the barrier system. Chemical emulsion should be applied with a suitable compressed air sprayer or power sprayer to facilitate uniform dispersal of the chemical emulsion. After the masonry foundation and the retaining wall of the basement come up, treatment of all backfill in immediate contact with the foundation structures should be done with chemicals at the rate mentioned on the label for the vertical surface of the substructure to a depth of 300 mm on each side of the structure. Earth filling below the floor should be levelled and treated before the sand bed or subgrade is laid. Special care should be taken to secure an intimate bond of poisoned soil at the junction of wall and floor. Treatments for masonry foundations with or without apron and for roller compacted concrete (RCC) foundations may be carried out.

7 Sporadic Pests and Their Control

7.1 Drywood Termite

Drywood termites are termites living in moistureless wood with no connection to the soil. They form nests in the structure or wood that they are infesting, such as door and window frames, furniture, etc. They are characterized by the presence of pelletized droppings in the infested area (Figs 7.1 and 7.2). Soldiers have mandibles with distinct teeth (Figs 7.3 and 7.4). Also, most drywood termite soldiers and workers are larger than the soldiers and workers in subterranean termite colonies.

Infestation from drywood termites usually begins via aerial routes. A mating pair can fly and colonize any suitable area with wooden structures, plants or trees. Decorative wooden hanging or rooftop gardens can also be preferred by this species for the establishment of a colony. This makes any portion of the house constructed with wood very vulnerable to attack.

Like other termite species, drywood termites are small and soft-bodied insects. The colony consists of the king and queen, the reproductive, the supplementary reproductive, the workers and the soldiers. Only the reproductive has wings. Drywood termite colonies are, however, very small compared to subterranean termite colonies.

7.1.1 Drywood termite control

Fumigation is the best method to treat infested furniture or other movable household items. Items can be moved into a fumigating chamber to be fumigated. Infested wood can also be treated with insecticides as described below. However, this work has to be done very carefully and skilfully.

Drywood termites infest dry, undecayed wood. If the infestation is small, such as in window frames or furniture, control can be achieved by insecticide application.

- Small holes about 30 cm apart are drilled into the wooden structure to reach the tunnels made by termites within the wood.

- Then, a suitable insecticide (recommended for drywood as wood treatment) is injected. The chemical should be injected until it overflows from the holes. This allows the chemicals to reach all portions of the gallery and the colony.
- Once the application is completed, the infested areas are covered with plastic sheets and the edges secured tightly with tape. This prevents evaporation of the chemicals and also protects the human inhabitants from fumes and contamination.
- The sheets can be removed after 24 h and the holes plugged with wood putty or plastic wood.

7.2 Powderpost Beetle

Powderpost beetles are a group of wood boring beetles. The larvae of these beetles stay inside the wood. They feed on wood and eventually reduce it to fine powder. This group is represented by a number of species all having a characteristic larval/grub behaviour. They cut into hard and dry wood, tunnelling through timbers in successive generations until the exterior is completely reduced to fine packed powder. Small holes are visible externally (Fig. 7.5). Often the sizes of these holes can be used to determine the species of the infesting beetle. The pests are often responsible for destruction of timbers of buildings, log cabins, furniture, windows and doors including frames, cabinets and any material made of wood.

Life cycle

Powderpost beetle larvae have a long life cycle, often varying from few to many months. Their entire development takes place inside the wood. Once the development is complete the young adults emerge from their galleries/tunnels by making an exit hole. These are often termed as pinholes. Often this action leads to piles of powdery frass near each exit hole. These pinholes normally range in diameter from 0.79–3.2 mm, depending on the species

Fig. 7.1. Pelletized dropping – a characteristic of drywood termite infestation. (Author's photo.)

Fig. 7.3. A typical drywood termite species, *Cryptotermes brevis*. (Courtesy of Pest and Diseases Image Library, Bugwood.org.)

Fig. 7.2. A close up of the pellets. (Courtesy of Josielyn Trinidad.)

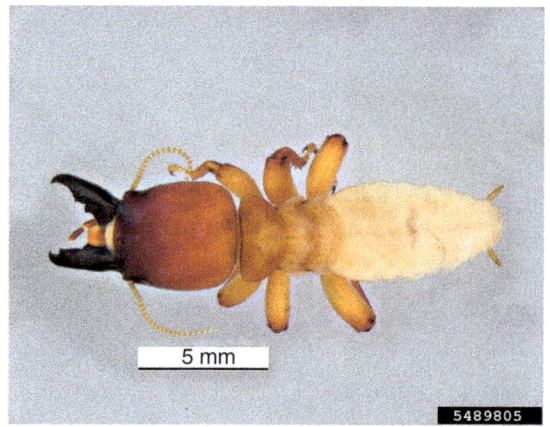

Fig. 7.4. Drywood termite soldiers have large mandibles (mouthparts) with teeth and their pronotum is as wide, or wider, than the head. (Courtesy of Pest and Diseases Image Library, Bugwood.org.)

of beetle (Koehler *et al.*, 2013). If wood conditions are right, emerging female beetles may mate, lay their eggs and re-infest the same wood, continuing the cycle for generations.

The most common types of powderpost beetles are Anobiid, Lyctid and Bostrichid beetles.

7.2.1 Anobiid powderpost beetles

Anobiid powderpost beetles are found infesting all types of structural as well as non-structural wood including various grades of plywood. The beetles prefer to infest wood with a high moisture content in poorly ventilated areas. Infestations are difficult to detect and the same wood can be re-infested year

after year. Emergence holes from Anobiids are 1.5–3.0 mm in diameter and round (Koehler *et al.*, 2013).

Appearance

Anobiid powderpost beetles are usually about 1.5–6.5 mm long and are reddish-brown or greyish brown to dark brown in colour. The body is cylindrical, elongated and covered with fine hair. The head is covered by the pronotum from the top view. The last three segments of the antennae are usually

Fig. 7.5. Fine wood powder being pushed from the exit holes of powderpost beetles. (Courtesy of Dr P. Suresh.)

Fig. 7.6. An adult Anobiid beetle. (Courtesy of Pest and Diseases Image Library, Bugwood.org.)

Fig. 7.7. Typical c-shaped larva of powderpost beetle. (Courtesy of Pest and Diseases Image Library, Bugwood.org.)

lengthened and broadened (Fig. 7.6). The larvae are white, c-shaped grubs with rows of small spines on the dorsal side (Fig. 7.7). The beetle life cycle lasts from a few months to a few years, depending on the species (Koehler *et al.*, 2013).

7.2.2 Bostrichid (false) powderpost beetles

Bostrichid powderpost beetles are 2.0–24.0 mm in length and are reddish-brown to black in colour.

Their bodies are elongated and cylindrical with a roughened thorax. Heads are concealed by the pronotum from above. The antennae are short with three enlarged, sawtoothed terminal segments (Fig. 7.8). The larva is white and c-shaped with no spines on the body. The life cycle lasts a year. Bostrichids infest all types of wood including bamboo items (Koehler *et al.*, 2013).

The emerging beetles leave round emergence holes measuring 3.0–4.5 mm in diameter. Sawdust-like frass sticks together in the emergence holes and is found tightly packed in galleries, but not in entrance holes (Koehler *et al.*, 2013).

Fig. 7.8. An adult Bostrichid beetle. (Courtesy of Pest and Diseases Image Library, Bugwood.org.)

7.2.3 Lyctid (true) powderpost beetles

Lyctid powderpost beetles are about 6.5 mm in length and brown in colour. The body is elongated and slightly flattened. The head is prominent and not covered by the pronotum. Antennae have a two-segmented terminal club (Fig. 7.9). The larvae are white and c-shaped, with the eighth abdominal spiracle enlarged. This beetle's life cycle is between 6 months and 4 years (Koehler *et al.*, 2013).

Fig. 7.9. An adult Lyctid beetle. (Courtesy of Pest and Diseases Image Library, Bugwood.org.)

Lyctids infest wood panelling, moulding, window and door frames, hardwood floors and furniture. Lyctids rarely infest wood older than 5 years. Therefore, infestations are usually in new homes or newly manufactured articles. Infestation usually results from wood that contained eggs or larvae at the time of purchase (Koehler *et al.*, 2013).

7.2.4 Control and management of powderpost beetles

Powderpost beetles are difficult to control. Their life cycle mostly takes place in a concealed area, which is difficult to treat. The best recommendation is always to use pre-treated wood for all types of construction. This prevents the occurrence of new infestations. In the absence of ready-to-use treated wood, it is advisable to treat the wood with wood treatment insecticides as a preventive measure. Coating finished wood with paints containing insecticides is another way to prevent infestation.

Powderpost beetles usually are known to lay eggs on untreated or unfinished wood. They avoid wood that is painted, varnished, waxed or similarly sealed. Beetles emerging from painted or varnished wood were either in the wood before finishing or were a result of re-infestation by eggs that were laid in emergence holes of adult beetles. Sealing holes prevents re-infestation from eggs laid within the hole (Koehler *et al.*, 2013).

The following methods reported by Koehler *et al.* (2013) have been found suitable and are recommended for prevention as well as treatment of infestations.

1. *Surface treatment* with insecticides labelled for surface treatment of bare, exposed wood is recommended. Spraying or brushing insecticides onto infested wood creates a barrier that kills adult beetles as they chew their way out of the wood. The barrier also kills newly hatched larvae as they attempt to bore into the wood. For the surface treatments to work properly, they must penetrate the wood. Therefore, the wood should be unfinished or sanded to remove the finish. In certain situations, the surface treatment can penetrate the wood sufficiently to kill larvae within the wood to prevent the further marring of the surface by additional emergence holes of adults.

2. *Fumigation* is considered the most effective method of controlling wood boring beetles. However, fumigation can be a costly method of control and does not provide residual protection of the wood. Movable items infested with powderpost beetles such as furniture, picture frames, etc. can be removed from the structure and fumigated inside a fumigation chamber.

7.3   Ant

Ants are sporadic pests and are often encountered in structures as a nuisance (Table 7.1). Ants are found foraging or nesting inside structures. Often ant reproductives are confused with swarming termites but they are distinguished as ant reproductives by having different sized fore and hind wing pairs, while in termites both pairs of wings are the same size.

Foraging

Ants have the habit of eating a wide variety of foods, including dead insects, seeds, nectar, meats, greases, sugars, fats and oils. Some ant species appear to just wander randomly; others trail one another precisely from colony to food source and back. Workers foraging for food communicate their messages through chemical signals creating trails inviting colony members to join them.

Life cycle

Male and female ants have wings and mate during the swarming flight (Fig. 7.10a). The male dies thereafter, while the female digs in the soil or wall voids or cracks in structures, thus making a nest and laying eggs. Eggs hatch into tiny, white, legless grubs (larvae), which are fed with salivary secretions from the female's stored body fats.

After several moults, the larvae change into soft, white pupae that look like motionless, white adults (Fig. 7.10b). Before they pupate, the larvae of some ants (carpenter ants and others) spin a silk cocoon. On hatching from pupae the emerging adults take on one of three roles or castes in the community: workers (all female workers and soldiers), female reproductive (future queens) or male reproductive.

7.3.1   Types of pest ants

Crazy ant (Paratrechina longicornis)

The ant can be identified by very long antennae and legs in comparison to the body, and the antennae are 12 segmented without clubs. They are small black coloured ants. The thorax lacks spines and the body profile is evenly rounded with a one-segmented pedicel (Fig. 7.11). Crazy ants are omnivorous and feed on insects, grease and sweet substances, but choice of food can change.

The crazy ant is an agricultural and household pest in most tropical and subtropical regions and in some temperate areas. They are common invaders

Table 7.1. Species of common ant pests and their characteristics. (Courtesy of Dr Andrew Giger.)

Species	Max. colony size	Nesting sites	Trails	Food preferences
Big-headed ants	300–3000	Usually in soil outside. Occasionally in crawl spaces or termite-damaged wood	Yes	Seeds and insects. Indoors: meat, grease, liver, molasses, peanut butter, fruit juices
Carpenter ants	15,000–100,000	Inside wood (preferably softened by fungus), in insulation, wall voids	Yes	Honeydew, plant and fruit juices, insects. Indoors: sweets, eggs, meats, cakes, grease
Crazy ant	2000	Under floors, in wall voids	Yes	Insects, sweets, other household food
Ghost ant	Thousands	Wall voids, behind baseboards, between cabinets and walls, soil of potted plants	Yes	Honeydew, dead and living insects, sweets
Little fire ant		Clothing, beds, food	Yes	Grease, fatty meat, peanut butter, oil, milk, fruit juices
Odorous house ant	10,000	Wall voids, crevices	Yes	Prefer sweets, but also grease, meat, cheese
Pharaoh ant	up to 100,000	Wall voids, behind baseboards, in furniture, under floor, between linens	Yes	Syrups, fruit, pies, meat, dead insects

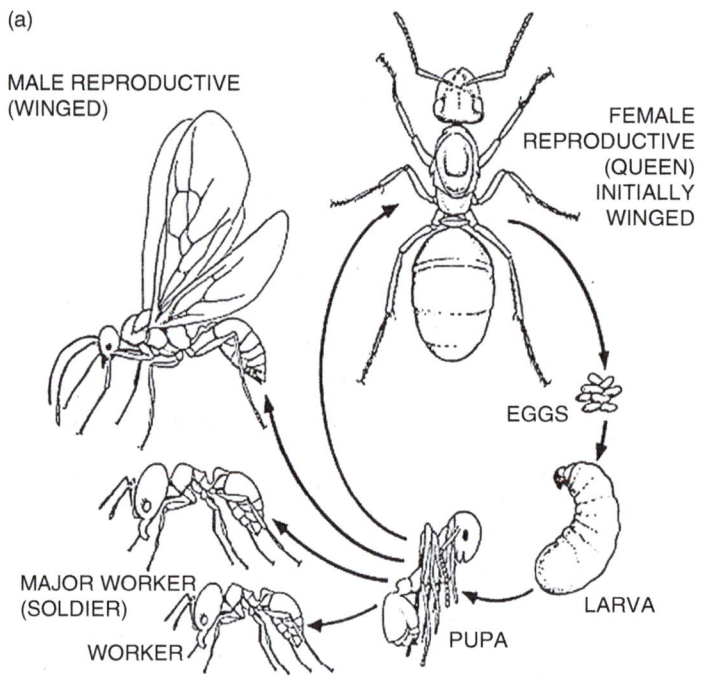

Fig. 7.10. (a) Typical life cycle of ants.

Fig. 7.10. (b) Typical ant life cycle stages showing eggs, larvae, pupae and adult. (Courtesy of David Cappaert, Bugwood.org.)

Fig. 7.11. Typical crazy ant. (Courtesy of Pest and Diseases Image Library, Bugwood.org.)

of structures and buildings. The ant gets its name from its habit of running about very erratically with no apparent sense of direction. Crazy ants can temporarily shelter in any manmade structure or items such as air conditioners, electrical boxes, cracks on the wall, in fact any place found to be dark, moist and warm. The colonies of these ants are often in soil, such as in planter areas, under logs, stones, landscape timbers, trash, wood debris, etc.

Pharaoh ants (Monomorium pharaonis)

The ant can be identified by usually pale and varying colour from yellowish to reddish with a darker abdomen. The antennae are 12 segmented with a three-segmented club, the thorax lacks spines and the pedicel is two-segmented (Fig. 7.12).

Pharaoh ants nest in wall voids, behind baseboards, under floors, inside furniture and between linens. They prefer to feed on syrups, fruit, stored items, meat and dead insects. Colonies can contain several thousand ants.

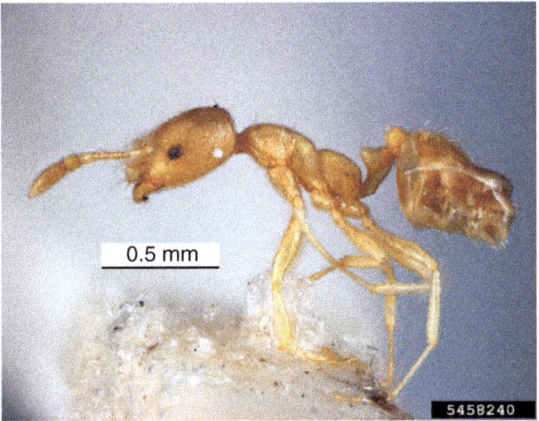

Fig. 7.12. Typical pharaoh ant. (Courtesy of Pest and Diseases Image Library, Bugwood.org.)

Ghost ant (Tapinoma melanocephalum)

The ant can be identified by its dark head and thorax compared to other body parts, the thorax lacks spines, the profile is evenly rounded with a one-segmented pedicel that is concealed from view from the top (Fig. 7.13).

Fig. 7.13. Typical ghost ant. (Courtesy of Eli Sarnat: Invasive Ants of the Pacific Islands, USDA APHIS ITP, Bugwood.org.)

In structures ghost ants nest in wall voids, behind baseboards, between cabinets and walls, and in soil. Their food includes honeydew, dead and live insects, and sweet substances. These ants can only survive in a moist environment.

Big-headed ant (Pheidole megacephala)

The ant can be identified by the major workers having exceptionally large heads in relation to their bodies. The ant thorax has two spines on the upper back, and the pedicel is two-segmented (Fig. 7.14).

These ants nest in soil, crawl spaces and damaged wood. Food includes seeds, insects, meat, grease, sweets and fruit juices.

Fig. 7.14. Typical big-headed ant. (Courtesy of Pest and Diseases Image Library, Bugwood.org.)

Little fire ant (Wasmannia auropunctata)

The ant can be identified by its light to golden brown colour with a slightly darker abdomen, the antennae are 11-segmented with clubs. The thorax has one pair of spines and the pedicel is two-segmented (Fig. 7.15). Little fire ants are known to inflict bites. They are particularly attracted to soiled clothing and they can chew clothes.

Little fire ants are frequent invaders of buildings, but they are generally known to be outdoor ants. Little fire ants can temporarily shelter in any man-made objects or items such as wall voids, air conditioners, kitchen cabinets and, in fact, any place found with moisture and warmth. They usually move indoors and nest in structures in heavy rain and hot weather.

Little fire ants are basically scavengers, looking for leftover food, food ingredients and similar objects to feed upon. Indoors they feed on oily substances and protein. Little fire ants basically enter the building through landscaping. They can also move into new structures from adjacent infestations, for example, from next-door apartments or elsewhere in the building itself. Once they move into new territory they quickly establish a colony and nest.

Fig. 7.15. Typical little fire ant. (Courtesy of Pest and Diseases Image Library, Bugwood.org.)

The Singapore ant (Monomorium destructor)

The ant can be identified by its near-uniform light yellow to dull brownish yellow colour with a darker gaster (swollen part of abdomen) (Fig. 7.16). The ant can be identified by the length of workers, which is highly variable (polymorphic) from 1.8–3.5 mm. The head and body are mostly smooth, shining and unsculptured except on the very top of the head, which has fine transverse ridges (which are inconspicuous). The antennae have 12 segments, including a three-segmented club (GISD, 2017).

The Singapore ant is described as a tramp ant as it is renowned for transporting itself around the world via human commerce and trade. The ant is known to nest both indoors and outdoors causing extensive economic damage in urban environments by gnawing holes in fabric and rubber goods, removing rubber insulation from electric and phone lines and damaging polyethylene cable (GISD, 2017).

Red imported fire ant (RIFA – Solenopsis invicta)

The ant can be identified by the pedicel, which consists of two segments. Workers vary in size (polymorphic) between 2 and 6 mm. The mandible has four distinct teeth and the antennae are ten-segmented, ending in a two-segmented club. A sting is present at the tip of the gaster. They are coppery brown in colour on the head and body, with a darker abdomen Fig. 7.17.

RIFA is a notorious invasive species known to be distributed through the world. The ant mound or nest has no obvious entry or exit holes. When the mound is disturbed, hundreds of reddish-brown worker ants crawl out in defence.

RIFAs are found in urban areas, especially in backyards, golf courses, parks, recreational areas,

Fig. 7.16. Singapore ant. (Courtesy of Pest and Diseases Image Library, Bugwood.org.)

Fig. 7.17. Typical red imported fire ant. (Courtesy of Pest and Diseases Image Library, Bugwood.org.)

school grounds, etc. If they enter structures or properties they are capable of harming the inhabitants. Nests can be built under pavements or even roads, as well as under driveways, foundations, lawns, edges of pavements (sidewalks), under patio slabs, in electrical boxes or near power lines. A colony can excavate huge quantities of soil, resulting in structural problems in driveways, paving and walls, and can also cause the formation of potholes in roads. Additional damage by mounds can be inflicted on trees, yard plants and pipes, including indoor and outdoor equipment and infrastructure.

Carpenter ant (Camponotus species)

These are large ants, measuring 0.76–2.54 cm, inhabiting many parts of the world. Carpenter ants can be identified by the general presence of one upward protruding node, looking like a spike, at the 'waist' attachment between the thorax and abdomen (petiole) (Fig. 7.18).

Carpenter ants are considered pests of dead and damp wood as they generally nest in them by making galleries chewed out with their mandibles. However, unlike termites they do not consume wood.

7.3.2 Control and management of common urban ants

Inspection

Inspection is critical in ant management. It is important to determine whether the ant colony is located inside or outside the structure. Looking for

Fig. 7.18. Typical carpenter ant. (Courtesy of Pest and Diseases Image Library, Bugwood.org.)

the following to determine the characteristics of the colony and the best control method is important:

- the species of ant;
- the full length of the trail;
- entry point of the ants into the structure;
- number of ant workers at a given time;
- if ants are converging into a specific place or passing by;
- if ants are carrying back food;
- if ants are carrying eggs or nymphs.

Habitat modification

Once the source of the infestation is located, the following methods should be followed to modify the environment:

- Determine whether ants are feeding or just passing from one site to another. Ants sometimes transfer shelter and this act may last a few days and subside naturally.
- Use habitat alteration to block ant entry points or to make the environment unfavourable.
- Caulk wall and sealing penetrations and mortar masonry cracks such as wall penetrations, including utility lines, air conditioning refrigerant pipes, phone lines, etc. Care should be taken to consider using sealants that cannot be cut or damaged by ants.
- Seal door and window frames.
- Remove water leaks.
- Store food sources in airtight jars.
- Follow up on sanitation with regular cleaning (counters, floors, kitchen appliances).

Recommended insecticide treatment

1. *Insecticide treatment of nests* is recommended when harbourage or a nest is detected. The nest should be thoroughly treated with the recommended dosage as per label directions. For example, imidacloprid-based formulations have shown good results.

2. *Food baits with insecticides* are the best way to control ants. Baits are available in granular and gel forms. However, the choice has to be based on the environment and the nature of the ant. Product labels have to be read carefully for the composition and type of ants mentioned before use. Ants can change their preference based on season, life cycle and environment, so at times it is best to choose a bait that is universal.

Baits are excellent in sensitive areas (e.g. electronic equipment, food counters or hospitals) where pesticide sprays are not appropriate. Baits are always applied inside bait stations or bait containers and secured to a surface using tapes or glue (Fig. 7.19). It is also advisable to mix the bait with the food the ants are commonly feeding on in the house to get the best result. (More about ant baiting is discussed on p. 99).

3. *Residual pesticide* should be selected for the treatment of cracks and crevices. Spray in areas where nests are suspected, as well as on pathways where they are most active. Dust formulations can be applied in wall voids. Aerosol formulations also work well to treat deep cracks and crevices and flush out infestations. Horticultural formulations should be selected for spraying in gardens and plants around the house in case ants are coming from an outdoor source.

When large ant hills or mounds are detected around the structure they can be treated with a recommended insecticide. Insecticide can be injected into the mound to saturate the site and kill or disable the colony.

Fig. 7.19. Gel baits are the best way to control large and small ant populations in structures. (Courtesy of Josielyn Trinidad.)

7.4 Psocid

Psocids (commonly referred to as booklice) are associated with microscopic moulds or fungus that grow on fabrics, clothing, carpets, curtains, wallpaper, panelling and walls, and in fact in any item that can retain moisture for a long time (Fig. 7.20). All the above-mentioned items can sustain growth of microscopic moulds or fungi in the presence of moisture. To the naked eye an area might look clean, but a microscopic examination might reveal the presence of moulds. These moulds serve as ready food for the nymphal and adult psocids.

Psocids are wingless forms but some species do have wings and always prefer shelter away from light sources. These insects are around 1.5 mm in length. These insects do not bring about any direct damage to humans, as they chiefly feed on mould attached to moist items, but sometimes they can extend their feeding habits to starchy materials like gum and starch pastes associated with book bindings, wallpaper and glued items used in furniture. These insects do not bite human subjects. Generally, psocids are classified under annoying pests and are not known to cause any type of damage.

Control and management of psocids

Psocids are controlled by reducing the moisture content in the room by proper ventilation and by

Fig. 7.20. Typical specimen of a psocid. (Author's photo.)

providing aeration and light. They can also be physically removed by using vacuum cleaners when present.

7.5 Flea

Fleas are occasional pests of homes and buildings that harbour pets and are accessible to stray animals. Fleas are small, wingless bloodsucking insects. Adult fleas are 1–4 mm long and have flat, narrow bodies. They are wingless with well-developed legs adapted for jumping. They vary in colour from light to dark brown. They feed mainly on the blood of mammals and birds. Of 3000 species only a dozen are known to be pests. The most important species are the rat flea, the dog flea and the cat flea. Their bites cause irritation, serious discomfort and loss of blood. The rat flea is important as a vector of bubonic plague and flea-borne typhus. The cat flea incidentally transmits tapeworms among cats and dogs (Fig. 7.21).

Behaviour

Fleas are found among hairs or feathers of pet animals such as cats and dogs. They are also found along with beds and clothing. Fleas avoid light. Fleas move around by jumping, and can jump as high as 30 cm.

Life cycle

Fleas shelter in close proximity to the host. The resting and sleeping places of the host are the usual

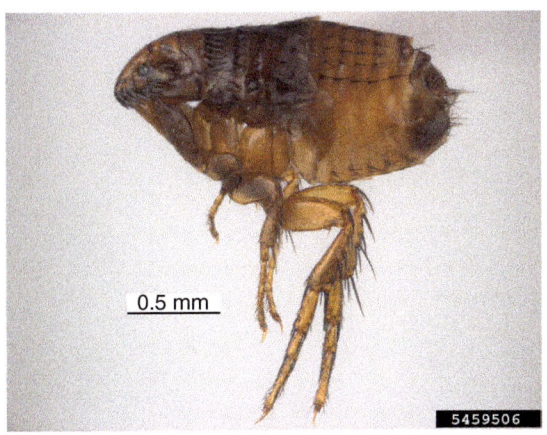

Fig. 7.21. Typical cat flea *Ctenocephalides felis*. (Courtesy of Pest and Diseases Image Library, Bugwood.org.)

breeding grounds for fleas. They can also be found with rubbish, in cracks in floors or walls, in carpets, pet shelter houses, animal burrows and bird nests. The life cycle of fleas passes through four stages, namely, egg, larva, pupa and adult. The eggs are deposited in cracks and crevices and the hatching larvae, measuring 4–10 mm long, are white in colour. They have no legs but are capable of movement. The larva spins a cocoon and pupates, usually hidden among house dust, dirt or in cracks. Under optimal conditions the development from egg to adult takes 2–3 weeks.

Fleas feed several times during the day or night. Heavy infestation with fleas is recognized by marks on clothing and bedding caused by undigested blood ejected by the fleas. Most flea species feed on one or two host species, but in the absence of their normal host they feed on humans. Adult fleas can survive several months without food.

Control and management of fleas

Once areas of flea populations are identified, the sites should be thoroughly vacuumed. Treating the areas surrounding the pest activity with a recommended insecticide is advisable. Treatment should cover floors, cracks in walls, wall voids and any area suspected of flea activity.

Treatment of pets

Pets infested with fleas should be treated by a veterinarian. Owners should clean, sterilize or wash all bedding material and pet kennels once a week until the flea population is controlled.

7.6 Tick

Ticks are not insects and can be easily distinguished by the presence of four pairs of legs in the adult (Fig. 7.22). Their bodies also lack clear segmentation. Ticks are ectoparasites, generally sucking blood from animals and humans. They occur around the world and are important vectors of a large number of diseases. Among the best-known human diseases transmitted by ticks are tick-borne relapsing fever, Rocky Mountain spotted fever and Q fever, and they can cause great economic loss. Two major families of ticks can be distinguished; namely, the hard ticks comprising about 650 species, and soft ticks, comprising 150 species (Rozendaal, 1997).

Fig. 7.22. A typical specimen of an ixodid tick, *Amblyomma inornatum*. (Courtesy of Clemson University – USDA Cooperative Extension Slide Series, Bugwood.org.)

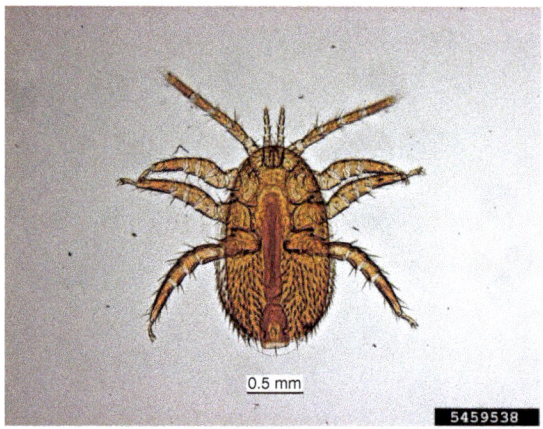

Fig. 7.23. A typical specimen of a parasitic mite. (Courtesy of Pest and Diseases Image Library, Bugwood.org.)

Life cycle

Ticks have a life cycle that includes a six-legged larval stage and one or more eight-legged nymphal stages. The immature stages resemble the adults and each of them needs a blood meal before it can develop into the next stage. Adult ticks live for several years, and in the absence of a blood meal can survive several years of starvation. Both sexes feed on blood (Rozendaal, 1997).

Prevention

The best way to prevent attack from ticks is to avoid infested places. If needed, precautionary measures should be taken such as use of body-covering clothing, use of repellents and avoiding walking through thick undergrowth. If ticks are spotted on the skin they should be immediately removed with care. All tick-related bites require a consultation with a physician.

7.7 Mite

Like ticks, mites are not insects and are occasional pests to humans. They have four pairs of legs and lack body segmentation (Fig. 7.23). Most species of mites are pests of agricultural crops but a few types of mites are parasitic on humans. In most species there are eggs, larval, nymphal and adult stages. The immature stages look similar to the adults but smaller.

Mites can present a serious biting nuisance to humans and animals. Many people show allergic reactions to mites or their bites. Some mite species cause a condition known as scabies. Mites are also vectors for rickettsial diseases, such as typhus fever due to *Rickettsia tsutsugamushi* (scrub typhus). The major mite pests of human interest are:

- biting mites (vectors of scrub typhus);
- scabies mites
- house dust mites.

7.7.1 Scabies mites

The sarcoptic itch mites, *Sarcoptes scabei*, infest the skin of a variety of animals including humans and can exchange hosts to some degree. Human scabies mites are very small and are rarely seen. They commonly attack the thin skin between the fingers, the bend of the elbow and knee, the penis, breasts and the shoulder blades. The mites burrow into the skin, making tunnels up to 3 mm long. When they first burrow into the skin, the mites cause little irritation, but after about a month, sensitization begins. A rash appears in the area of the burrows and intense itching is experienced (Potter, 1996).

Scabies mites are transmitted by close personal contact, usually from sleeping in the same bed. Bedridden individuals in institutions (e.g. nursing homes) may also pass the mites from caregiver to patient. The adult fertilized female mite is usually the infective life stage. The female adheres to the skin using suckers on its legs and burrows into the skin, where she lays her oval eggs. In 3–5 days these eggs hatch into larvae and move freely over the skin. Soon they transform into nymphs and reach maturity 10–14 days after hatching (Potter, 1996).

Control and management of mites

A scabies infestation should be treated as a medical problem. An evening soap bath followed by overnight treatment is recommended. This gives the pesticide medication 8–12 h to work. Commonly used products include permethrin and crotamiton. Reading the directions on the product label is a must before use (Potter, 1996). Scabies mites cannot live without a human host for more than 24 h, so an insecticide treatment of the house is not required. It is, however, recommended that all personal items such as clothing and bedding from infested individuals be washed in hot water or dry cleaned.

7.8 Sandfly

Sandflies are small bloodsucking flies (Fig. 7.24). They are important as vectors of a disease called leishmaniasis. Leishmaniasis is transmitted by the bite of infected female sandflies belonging to the species *Phlebotomine*. This species can transmit the protozoa *Leishmania*, the disease-causing organism, to humans. The disease has different forms. The cutaneous form manifests with skin ulcers, while the mucocutaneous form leads to ulcers of the skin, mouth and nose, and the visceral form starts with skin ulcers and then later presents with fever, low red blood cells, and enlarged spleen and liver.

Species that occur in the Mediterranean region can spread sandfly fever, a viral disease also known as 'Pappataci fever' or 'three day fever'.

Life cycle

Sandflies are an occasional pest notable in some parts of Asia. They are mostly reported from rural, semi-urban and seashore areas. Female sandflies deposit their eggs in humid places on damp soil or wet sand rich in humus or manure. The larvae feed on decaying organic matter. Examples of suitable breeding sites are small cracks and holes in the ground, the ventilation shafts of termite hills, animal burrows, cracks in mud walls and masonry, tree roots and shaded sandy beaches (Fig. 7.25). Large populations of sandflies can build up in areas like cattle sheds and animal farms. The cattle can provide an abundant source of blood, while the stables and houses provide suitable resting places. The life cycle is generally completed in 1–4 months depending on the climate and food availability.

Behaviour

The adult sandflies are weak fliers and usually stay within a few hundred metres of their breeding places. They move in a characteristic hopping style, with many short flights and landings. Most biting occurs outdoors, but a few species are also known to move inside homes and may bite indoors. Most species are active at dawn and dusk and during the night, but in forests and indoors they may also attack in daytime, especially if disturbed by human activities.

Sandflies feed on plant juice. Females need a blood meal in order to develop eggs. Blood is taken from humans, and animals such as dogs, farm livestock, wild rodents, snakes, lizards and birds. Each sandfly species has specific preferences for its source of blood, but the availability of hosts is an important factor.

Control and management of sandflies

PERSONAL PROTECTION AGAINST SANDFLY BITES. The recommendations given out by the US Center for Disease Control (2013) as preventive measures for both outdoor and indoor conditions when travelling to infested areas are as follows:

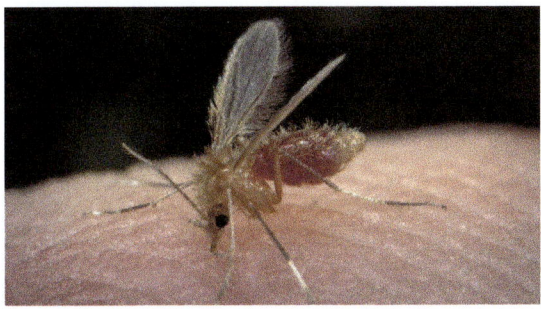

Fig. 7.24. Typical sandfly adult. (Courtesy of Muhammed Kufrevi.)

Fig. 7.25. Wet beach sand often acts as a breeding ground for sandflies in South-east Asia. (Author's photo.)

In outdoor situations:

- Minimize the amount of exposed (uncovered) skin. To the extent that is tolerable in the climate, wear long-sleeved shirts, long trousers and socks, and tuck your shirt into your trousers. (See below about wearing insecticide-treated clothing.)
- Apply insect repellent to exposed skin and under the ends of sleeves and trouser legs. Follow the instructions on the label of the repellent. The most effective repellents generally are those that contain the chemical DEET (N,N-diethylmetatoluamide).

In indoor situations:

- Stay in well-screened or air-conditioned areas.
- Keep in mind that sandflies are much smaller than mosquitoes and therefore can get through smaller holes.
- Spray living/sleeping areas with an insecticide to kill insects.

- Use a bed net. If possible, use a bed net that has been soaked in or sprayed with a pyrethroid-containing insecticide. The same treatment can be applied to screens, curtains, sheets and clothing (clothing should be retreated after five washings).

RECOMMENDED INSECTICIDE TREATMENT. Environmental control efforts are the best way to check the breeding and harbourage. Removing moist soil, organic matter, animal waste and other suspected items from the vicinity is a must.

1. *Insecticide application*: Control of sandflies is often difficult because methods mainly rely on interrupting contact between female flies and humans. Both chemical and environmental control are the best methods to control the population of sandflies. The main chemical control methods are indoor residual spraying with organophosphates (malathion),

carbamates (propoxur) and synthetic pyrethroids (permethrin and deltamethrin). Chemical control may have challenges if the fly population has developed resistance. The combination of a number of active agents and chemical rotation is thus recommended.

2. *Bait application*: Use of a product called 'attractive toxic sugar baits' (ATSBs) has shown success in controlling sandfly populations. This product is applied in patches of vegetation and barrier fences in areas lacking vegetation that could be sprayed. This method has been successful because both male and female sandflies, like other biting flies, require sugar from plants and sometimes honeydew for survival.

7.9   Wasp

Wasps are sporadic pests on structures, often noticeable by their nests (Fig. 7.26). Paper wasps, for example, tend to build small, umbrella-shaped nests under eaves and overhangs. Hornets build large nests shaped like a football. The nests are constructed using raw wood converted into paper pulp. The nests usually occupy the corners and angles between walls and window or roof corners. These nests are constructed at a certain height for protection.

Fig. 7.26. Typical paper wasp, *Ropalidia socialistica*. (Courtesy of Pest and Diseases Image Library, Bugwood.org.)

Wasps are mostly predatory in behaviour, constantly on the lookout for other prey insects and spiders. They act as beneficial insects removing other insect pests from plants and areas of harbourage. They are either solitary or live in social colonies like hornets. They are not known to carry any diseases and are generally harmless.

Controlling them is purely by physical removal and relocation, handled by an expert.

8 Stored Product Pests

Stored product insects comprise only two insect groups or orders. These insect orders are moths (Lepidoptera) and beetles (Coleoptera). These are further classified as either:

- *Internal feeders*: Internal feeders of stored products feed internally, causing damage in the grain kernel as grub or larvae. For example, grain weevils and grain borers.
- *External feeders*: External feeders of stored products feed on grain that has been damaged or milled during processing. Damaged grain kernels have exposed endosperm that is accessible food for insects and fungi. For example, Indian meal moths.

Depending on the type of food processed and type of stored ingredients, the risk of internal or external feeders can easily be identified. For example, bakeries predominantly use milled flour and other milled ingredients; these face damage from external feeders. Breweries or snack food manufacturers that store whole grain or kernel ingredients mostly have a risk of internal feeders (Figs 8.1 and 8.2).

Fig. 8.2. Cereals are the most common product to suffer damage from stored product pests. (Courtesy of Clemson University – USDA Cooperative Extension Slide Series, Bugwood.org.)

Fig. 8.1. Expensive commodities such as nuts, dry fruits and spices are often damaged by stored product pests. (Author's photo.)

Some common stored product pests are discussed below.

8.1 Rice Weevil (*Sitophilus oryzae* and *S. granarius*)

The weevil is dark brown in colour, measuring 4 mm in length (Figs 8.3 and 8.4). After mating the female beetle makes a small hole in the grain, deposits an egg and covers it with a gelatinous fluid. A female will lay up to 400 eggs. The legless grub feeds inside the grain once developed, pupates and emerges through an irregular hole made in the grain. The life cycle ranges from 26–28 days. The insect infests the grain both in storage and in field. It is destructive to wheat, corn, rice, etc.

Fig. 8.3. *Sitophilus oryzae*: Body length 2.3–3.5 mm, similar in appearance to the granary weevil, however, the rice weevil is reddish-brown to black in colour with four light yellow or reddish spots on the corners of the elytra; prothorax is strongly pitted and the elytra have rows of pits within longitudinal grooves. (Courtesy of Pest and Diseases Image Library, Bugwood.org.)

Fig. 8.4. *Sitophilus granarius*: Body length 3–5 mm; dark brown to nearly black in colour, ridged elytra. Pronotum almost as long as elytra. (Courtesy of Pest and Diseases Image Library, Bugwood.org.)

8.2 Sawtoothed Grain Beetle (*Oryzaephilus surinamensis*)

The sawtoothed grain beetle is a common grain pest (Fig. 8.5). The species has characteristic sawtooth like projections on each side of the adult thorax. It feeds on a wide range of milled cereals, dried fruits, candies and nuts.

Fig. 8.5. Sawtooth grain borer. (Courtesy of Clemson University – USDA Cooperative Extension Slide Series, Bugwood.org.)

The life cycle of the beetle is around 3–5 weeks. The female lays up to 400 eggs, which are laid loosely on the surface of the grains. The larvae develop very quickly if presented with high temperature and moisture. Adults are known to live up to 3 years.

8.3 Red Flour Beetle (*Tribolium castaneum*) and Confused Flour Beetle (*Tribolium confusum*)

Flour beetles are common pests in stored products across the world. The length of the adult beetle is 3–4 mm, and it has a reddish-brown body colour. The beetle is a generalist feeder, preferring milled grains, seeds, vegetable powders, dry fruits, oilcakes, nuts and museum specimens like dry insects and stuffed material. These insects are known to produce toxic quinone compounds, which gives a distinct foul odour and taste to the food products that they infest.

Flour beetles are two species consisting of red flour beetles and confused flour beetles. The antennae of the confused flour beetle gradually expand towards the end, while those of the red flour beetle abruptly expand at the end to form a club of three segments. The red flour beetle is capable of flight but the confused flour beetle does not fly. The adults are very active, especially in the evening hours and are occasionally trapped in light traps (Figs 8.6 and 8.7).

The life cycle of these beetles is approximately 3 weeks but is dependent on the temperature.

Females can lay up to 450 eggs. These are all laid on the surface of grains. Hatching larvae develop into adults in 4–7 weeks under optimal conditions. Adults can live for a few months to 2 years.

8.4 Cigarette or Tobacco Beetle (*Lasioderma serricorne*)

The cigarette beetle is a pest of all forms of tobacco, including refined cigarette packets as well as of stored tobacco leaves as bales. It is also a minor pest of oilcakes, oilseeds, cereals, dried fruit, sage, flour and some animal products. The beetle is light brown in colour with a round look, with its thorax and head bent downward and this presents a strongly humped appearance. Antennae are of uniform thickness (Figs 8.8 and 8.9).

Fig. 8.6. Typical confused flour beetle. (Courtesy of Clemson University – USDA Cooperative Extension Slide Series, Bugwood.org.)

Fig. 8.8. Typical adult cigarette beetle. (Courtesy of Pest and Diseases Image Library, Bugwood.org.)

Fig. 8.7. Typical red flour beetle. (Courtesy of Clemson University – USDA Cooperative Extension Slide Series, Bugwood.org.)

Fig. 8.9. Larva of the cigarette beetle. (Courtesy of Pest and Diseases Image Library, Bugwood.org.)

The female beetle lays around 100 eggs loosely on the surface of a suitable substrate. The larvae, once hatched, are active and they generally bore into the substrate by feeding on it. The complete life cycle takes around 3 weeks at 37°C and 4 months at 20°C.

8.5 Drug Store Beetle (*Stegobium paniceum*)

The drugstore beetle (*Stegobium paniceum*) is commonly known as the bread beetle or biscuit beetle. Adults can measure up to 3.5 mm, are brown in colour and can be found infesting a wide variety of dried plant products, especially stored food like biscuits and crackers. One way to distinguish them from cigarette beetles is that they have antennae ending in three-segmented clubs. The drugstore beetle also has grooves running longitudinally along the elytra, whereas in the case of cigarette beetle it is smooth (Fig. 8.10).

The female beetle lays eggs in batches. Hatched larvae tunnel into stored products and develop. The larval and pupal periods vary between a few months and nearly 2 years.

Fig. 8.10. Typical adult drug store beetle. (Courtesy of Pest and Diseases Image Library, Bugwood.org.)

8.6 Carpet Beetle (*Anthrenus* spp.)

Carpet beetles are occasional pests of stored product commodities. There are a number of species but the most frequent ones are *Anthrenus flavipes* and *A. verbasci*. Carpet beetle adults measure between 2.5 and 5.0 mm in length. The adult beetle is dull black in colour with the surface of the wings characteristically mottled with white, brown, yellow or black (Fig. 8.11). The beetles are known to fly and are attracted to light.

Carpet beetles feed on dead animal skins, bird feathers and all types of animal hairs. Items such as wool, felts and material containing them are very susceptible to carpet beetle attack, as well as stored cereals or dry pet food.

The female carpet beetle lays about 100 eggs. Eggs hatch in 2 weeks. The developmental time from egg to adult is approximately 3–4 months under optimal conditions.

Fig. 8.11. A typical adult carpet beetle. (Courtesy of Pest and Diseases Image Library, Bugwood.org.)

8.7 Indianmeal Moth (*Plodia interpunctella*)

This moth is distributed in a wide range of geographic conditions and is found in many types of food processing and storage facilities. The moth has brown forewings with a white band (Figs 8.12 and 8.13). It lays about 300 eggs in clusters and the life cycle is competed in about 5–6 weeks. The larvae feed on grains and dried fruits. The larvae are general feeders and the adults do not feed. The larvae produce a dense webbing. The life cycle of the moth is around 26 days under optimal conditions of humidity and temperature.

Fig. 8.12. Typical adult Indianmeal moth. (Courtesy of Pest and Diseases Image Library, Bugwood.org.)

Fig. 8.13. Adult Indianmeal moth with wings open. (Courtesy of Pest and Diseases Image Library, Bugwood.org.)

8.8 Control and Management of Stored Product Pests

Stored product pests in household items can be managed easily by following simple measures. Controlling the humidity and keeping the stored items as dry as possible in airtight containers is always recommended. Periodic checks and resealing helps longer storage. However, when it comes to control of these pests in commercial establishments such as warehouses, stores, mills or breakfast cereal manufacturing units, a whole new type of effort is needed. The tasks become complex and involve multi-level planning and management.

The following methods are recommended for management.

Inspection

Special attention is needed for products and items that have a long shelf life, such as dry animal food, beans, cereals, flour, as well as processed food made from cereals, such as biscuits and noodles. Storing the product in moistureless containers or sealed packaging is recommended. However, this is not a foolproof method as infestation might be carried in the item beforehand. Continuous inspection and monitoring can help early detection along with quick action to minimize the damage.

Habitat modification

Once an infestation is detected the damaged goods need to be quickly separated from the rest of the stock. An intensive cleaning of the store room and warehouses is a must. Cracks and crevices need to be surface treated with insecticide. Moisture control is a must in prevention of new infestation. Thoroughly dried products stored in airtight bags are the best way to prevent reoccurrence of pest attack.

8.8.1 Insecticide application

The best way to treat infested areas is to use registered pesticides for food areas and apply it to cracks and crevices, floors, pallets, bins, transport and transportation items. This should be done after removing the infested material from the vicinity. To achieve faster and more effective control in large commercial areas, fumigation is recommended. A fumigation expert needs to be consulted for this work.

Follow-up

Setting up of a regular monitoring programme in warehouses and food storage areas is a must. Pheromone traps for stored product beetles are very helpful in such programmes, including the use of light traps; yellow coloured glue board traps also help in pest capture.

8.8.2 Fumigant application

Fumigants are gases and are broad spectrum pesticides that can act as respiratory poisons, anaesthetics or as narcotics. As they are gaseous in nature and have acute inhalation toxicity, fumigant products are all labelled under Toxicity Category I with the signal word Danger.

There are different types of fumigants but the most common ones available for stored products are phosphine and methyl bromide. Sulfuryl fluoride has been recently registered as a fumigant for post-harvest use. Depending on the type and nature of infestation, heat can also be used in combination with other fumigants and pest control measures to increase efficiency of the treatment.

9 Vertebrate Pests and Their Control

9.1 Rats and Mice

Rats and mice are possibly the most noteworthy pests to humans. Rodents eat and contaminate food, damage structures and property, and transmit parasites and diseases to other animals and humans. They live and thrive in a wide variety of climates and conditions, and are often found in and around homes, buildings, gardens, various manmade structures, farms, open fields and forests. Their ability to transport themselves with humans and human-driven activities has thus made them a cosmopolitan pest.

Control and management of rodents remains predominately a challenge with a need for continuous monitoring and intervention. Only rat-proofing through physical methods has proven to be long-lasting. Other methods such as traps, baits and poisoning are labour-intensive, messy and, at times, dangerous. The need for alternative solutions is a priority for every manufacturer, researcher and pest controller.

Feeding and activity

Both rats and mice are active after dark. Feeding is a major night-time activity. Another activity that rats cannot live without is 'gnawing'. They are capable of chewing through building materials such as plaster block, wooden board, aluminium siding, brick, wall board, wooden cabinets and plastic. This activity often leads to severe property damage.

Rodents are capable of feats that make them a hard-to-control pest. Rodents have been known to compress their bodies and squeeze through an opening only 2 cm wide, making no place in a building safe. They are capable of jumps and leaps up to 1 m in the air, which makes most outdoor and indoor areas of a structure accessible to them. They can climb pipes and balance themselves on thin wires. Rodents can swim long distances, making their travel through most city sewers easy.

They have poor vision, which is well compensated for by their keen senses of smell and hearing. They use their whiskers and guard hairs to 'touch' their way through pathways. Rodents also have a keen sense of taste. They can detect food chemicals up to concentrations of parts per million, thus explaining why rats often reject baits or avoid traps.

Nesting behaviour

Rats make use of all available resources to make a shelter. Generally, outdoors they live in burrows dug up in gardens or on vacant land. Both roof rats and Norway rats nest indoors. The nests are usually among clutter, in wall gaps, ceilings, crawl spaces, underneath goods and equipment, inside cabinets, etc.

9.1.2 Types of pest rats and mice

There are only a few species of rats and mice that have attained pest status because of their ability to live with humans. They have varied life cycles (Table 9.1) and consist of the following:

- Norway rat;
- roof rat; and
- house mouse.

Norway rat

The Norway rat (*Rattus norvegicus*) is also called the brown rat, house rat and sewer rat. Adults are large, measuring around 40 cm and weighing an average of 500 g. One distinguishing characteristic of this rat is that the tail is shorter than the combined length of the head and body (Fig. 9.1).

In structures, regular household garbage attracts Norway rats. Once food is located they either eat the food on the spot or, at times, carry the food to another safer location to feed. Rats are known to collect and store food. They are in general neophobic:

avoiding new objects in their territory. These rats are omnivorous and will feed on almost anything available to them.

Roof rat

The roof rat (*Rattus rattus*) also called the black rat, can measure around 40 cm in length with a characteristic long tail, large ears and eyes, with a pointed nose. Unlike Norway rats the length of the tail is longer than the total length of the head and body (Fig. 9.2).

Roof rats are mostly found in the upper parts of structures. As their name suggests, roof rats are commonly spotted in elevated areas such as trees, rafters, attics, ceilings and roofs. Roof rats can also nest on the ground if necessary. These rats are omnivorous and will feed on almost anything available to them.

Table 9.1. Life cycle details of common pest rodents.

	Norway rat	Roof rat	House mouse
Gestation (days)	22–24	22–24	20–21
Litter size	6–12	6–12	4–7
Weaning (weeks)	3–4	3–4	3–4
Maturity (months)	3–4	2–3	2–3
Life span (years)	1–3	1–3	1

House mouse

The house mouse (*Mus musculus*) is a highly domesticated species (Fig. 9.3). Depending upon food supplies their habitats are limited and are usually within 10 m of their nesting place. These mice are omnivorous and will feed on almost anything available to them.

9.1.3 Control and management of rodents

Rodents must be understood before they can be controlled. A good knowledge of their behaviour, feeding range and other factors is essential for their management.

Inspection for activity

A good inspection can lead to the detection of many signature rodent activities that can help determine the species, the level of activity, the pathways and harbourage of the infesting rodents. This helps the practitioner choose various control measures and design a maintenance programme to put in place. The following items should be checked when inspection is carried out:

Rodent droppings: A single rodent may produce 20–30 droppings daily. Based on the type, size of the

Fig. 9.1. Typical Norway or sewer rat. (Author's picture.)

Fig. 9.2. Typical roof rat. (Author's picture.)

Fig. 9.3. Typical house mouse. (Author's picture.)

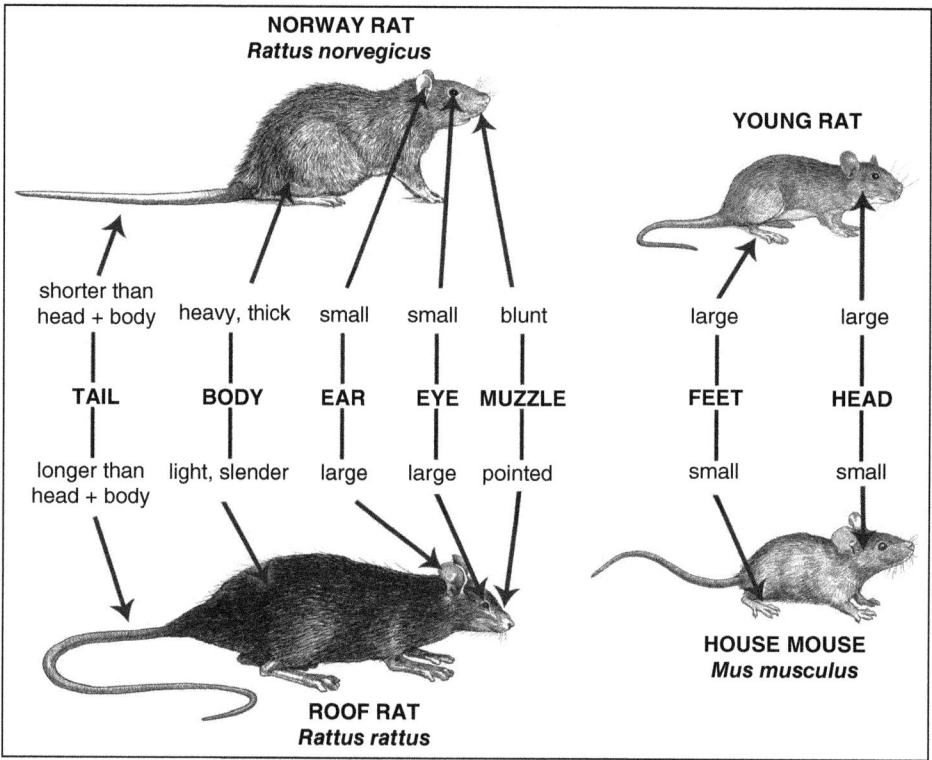

Fig. 9.4. A simple method of identifying common pest rodents. (Author's picture.)

dropping and the location of the rodent species, the number of rodents as well as the area of harbourage can be ascertained. New droppings can indicate the level of activity (Fig. 9.5).

Gnawing damage: Rodents need to gnaw to keep their incisor teeth in shape. Their teeth grow at a rate of about 2.5 cm per year. Rodents keep their teeth worn down by continuously working them against each other and by gnawing on hard surfaces. Indication of gnawing damage as evidence of

a rodent infestation will show rodent activity as well as paths taken.

Runways: Non-resident rodents, such as those living outdoors, constantly travel the same route. Their runways appear as beaten paths on the ground. Look next to walls, along fences, under bushes and buildings. Indoor runways sometimes also make clear trails by leaving hairs, foot marks and greasy marks on walls. These are clear signs of pathways, which can be used for tracking powder application.

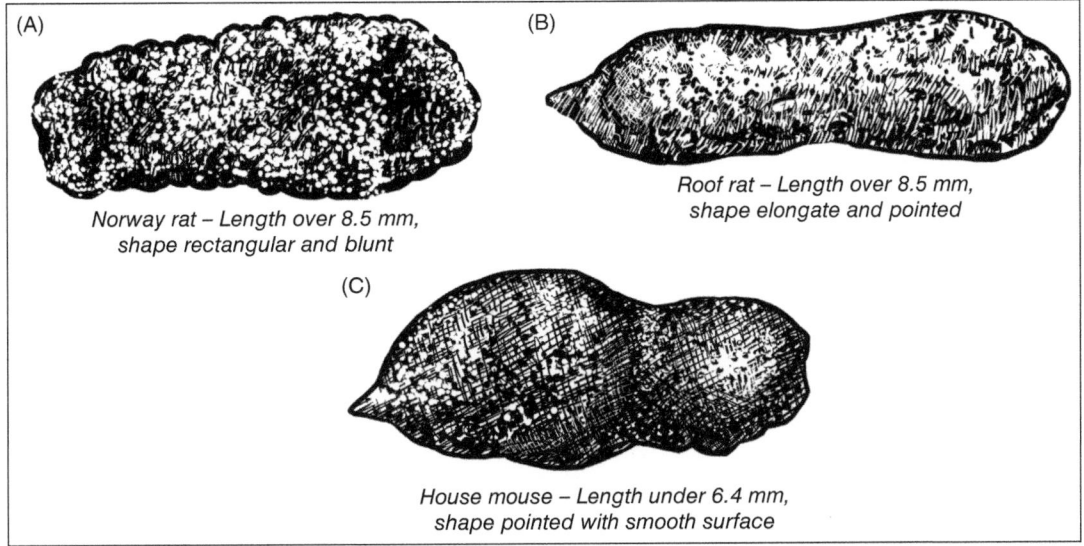

(A)

Norway rat – Length over 8.5 mm,
shape rectangular and blunt

(B)

Roof rat – Length over 8.5 mm,
shape elongate and pointed

(C)

House mouse – Length under 6.4 mm,
shape pointed with smooth surface

Fig. 9.5. Characteristics of rodent droppings, a tool for rodent identification.

Burrows: Outdoors, rodent burrows may be found singly or in groups along foundation walls, under slabs and dumpster pads, in overgrown weedy areas, beneath debris and in embankments.

Rodent sightings: First-hand rodent sightings from the inhabitants of the building, including the location and the direction of their travel or runs, are the most critical pieces of information needed to start an inspection. This information often leads to the discovery of harbourages, wall gaps and pathways, and preferred feeding spots.

Rodent-proofing and exclusion

A rodent control programme should start with rodent-proofing and exclusion. However, this is generally applicable if the rodent is a non-resident one, and is suitable for rodents visiting the structure for food and temporary shelter. For resident rodents, rodent-proofing works by restricting their movements and keeping them away from their regular food and water spots. Rodent-proofing causes rodents to accept baits far more readily as the area for rodent movement is reduced, which makes eradication quicker.

Rodent-proofing is achieved by checking throughout the following:

- perimeters of the structure such as the four walls and the roof and floor;

- doors and windows;
- utility pipes, drains, manholes and conduits connecting the structure to public services;
- gardens for burrows;
- adjacent structures, trees and paths as access ways; and
- incoming goods and people.

It is generally observed that most practitioners shy away from the above job scope as they often find themselves not qualified or lacking the training to carry out these evaluations. In such situations the practitioner needs to coordinate or integrate his or her management programme with the building engineers or in-house maintenance staff. If it is a residential property, he or she needs to bring an experienced person with him/her.

Rodent-proofing is done using any appropriate construction material, such as cement, metal sheets, metal mesh and sealants. Wood and plastic should be avoided as rodents can chew through these materials and will gain access over time. Various user-friendly rodent-proofing materials such as sealants, fast-setting cements and customized metal rolls are now available in the pest control market for easy handling and use as follows:

- Galvanized sheet metal, 24 gauge or heavier, is recommended for most general uses to exclude rodents.

- Cement patching powder is a material that has similar physical characteristics to cement. It is available in small-sized containers and is easy to mix. Most brands harden in less than 4 h and provide good to moderate rodent exclusion.
- Epoxy and fibreglass resins can be used as caulking and hole-filling materials. These materials are available from auto and boat repair supply sources; many formulations harden quickly and are very durable, weather resistant and rodent-proof.
- Butyl caulking is a material with slow curing. It is very good for sealing gaps between metal and masonry and for joints up to 2 cm wide and 1 cm deep.

Use of traps and monitors

Various types of rodent traps and monitors are available on the market for detection of rat and mouse activity as well as for control of low-level infestations.

RAT GLUE. Glue in the form of 'ready-to-use glue boards' or 'paint on glue in cans' is available on the market for use. Although most often used against mice, it is sometimes effective against rats. However, it has to be ensured that these conform to various standards of trapping. Poor quality or unspecified grades of glue will fail to trap rodents over a certain body weight and size.

It is also important to be aware that some people consider glue boards inhumane because the rodents are not killed instantly.

Glue board placement:

- Always place the glue boards in the same location as the snap traps.
- Secure the glue board with a nail or wire or double-sided tape so a rat is unable to drag it away.
- Place them on overhead runways along pipes, beams, rafters and ledges.
- Never place glue boards directly over food products or food preparation areas.
- Install glue boards inside bait stations in sensitive situations, with sensitive clients, in the presence of children or pets, or in areas with excessive dust or moisture.
- Check glue boards frequently and dispose of rodents humanely.
- Sometimes adding a bait in the centre of the glue board will improve its effectiveness.

SNAP TRAPS. Snap traps come in various types and sizes and are an effective method of killing rodents when used correctly (Fig. 9.6). Trapping is advised for use in places where rodenticides are considered too risky or aren't working well, in sensitive areas where visibility and odour of dead rodents may not be a concern or when rodent infestation is low and isolated. However, the quality of these traps should be checked before integrating them into

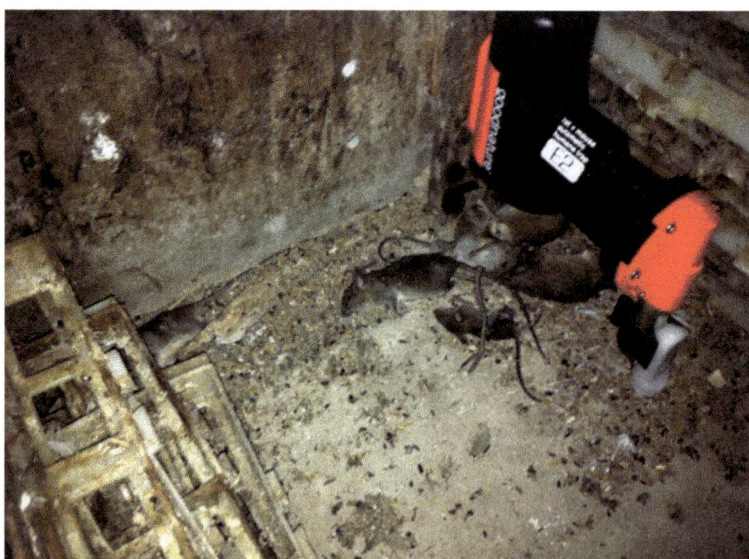

Fig. 9.6. A multi-shot snap trap capable of self-loading up to 24 times. (Courtesy of Ken Stern.)

pest control programmes. Traps unable to snap and kill instantly and correctly may pose concerns for clients.

CAGE TRAPS. Cage traps are traps used along with food baits without poison to trap rats and mice. They work well in isolated and low-level infestation situations. The biggest challenge is, however, to extract and cull the rodent humanely, which is a crucial job for the technician. Cage traps are useful when the species of rodent to be trapped is large. Cages need to be washed thoroughly and dried before being used to free the surfaces from the odour of previously trapped rodents. Lingering smells sometimes may reduce the efficacy of trapping. Various types of sprayable food flavours are available for use in cages. These flavours generally mask the odours left from previous trappings.

Rodent baiting

NEED FOR PRE-BAITING. Baits containing no active ingredient can be used to pre-bait rodents (Fig. 9.7). This helps to determine the location and quantity of eating activities, identify the species of rodent and also get them used to the bait. A suitable bait dye can be used in the bait formulation to help identify the correct droppings from pre-bait consumption (Fig. 9.8). This helps distinguish treated from stray droppings. Once data are collected, the pre-baits can be replaced by actual bait. Bait and pre-baits can be interchanged to get the best result and reduce bait shyness.

BAITING WITH METAL PHOSPHIDES. Pest rodents can be quickly controlled by using metal phosphides. These compounds are single-dose fast-acting rodenticides and death occurs within 1–3 days after a single bait ingestion. Bait is prepared by mixing a food substrate and a phosphide such as zinc phosphide. Zinc phosphide is typically added to a food bait at a concentration of 0.75–2.0%. The mixture is packed in a paper bag and placed in a sealed bait box. Bait boxes are positioned suitably in rodent active areas. Once consumed partly or fully, the stomach acids of the rodent react with the phosphide to generate the toxic phosphine gas, which kills the pest.

Zinc phosphide baits are cheaper than most second-generation anticoagulants on the market. In the case of a large rodent infestation, their population can be reduced by large amounts of zinc phosphide bait application. Generally, repeated applications of zinc phosphide on the same job site are not recommended. The introduction of anticoagulant bait as maintenance bait is advisable to overcome bait shyness as well as resistance development. These methods of alternating rodenticides with different modes of action increase the chances of eradication of the rodent population in the area. Also, metal phosphides do not accumulate in the tissues of poisoned animals, so the risk of secondary poisoning is low.

BAITING WITH ANTICOAGULANTS. Anticoagulant rodenticides work by stopping normal blood clotting. The compounds act by blocking the vitamin K cycle, resulting in an inability to produce essential

Fig. 9.7. A pre-bait pellet containing a dye and no poison. (Author's photo.)

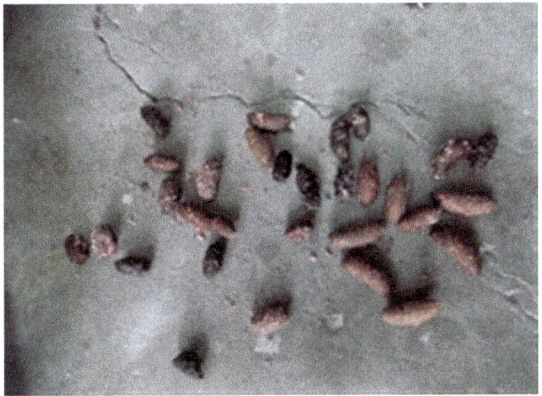

Fig. 9.8. Faecal pellets with the coloured dye, a sure indication of bait consumption. (Author's photo.)

blood-clotting factors – mainly coagulation factors II (prothrombin) and VII (proconvertin). This eventually results in the rodent bleeding to death internally.

There are a number of such compounds registered for use on the current market, such as bromadiolone, chlorophacinone, difethialone, brodifacoum and warfarin. Anticoagulants are defined as chronic, which means death occurs 1–2 weeks after ingestion of a dose. Most popular are the single-dose anticoagulants that are second-generation compounds, compared to the multiple-dose (first-generation) rodenticides. The main benefit of anticoagulants over other poisons is that the time taken for the poison to induce death is longer and rats do not associate death with the bait.

Rodenticide baits using anticoagulant are available in the form of pellets, blocks and paste. They can also attract other animals – particularly vulnerable are pets and wildlife. Thus, baiting has to be carefully planned and baits need to always be placed inside tamper-proof, sealed bait stations. The stations also need to be secured to the ground or wall so as not be removed by accident. A cautionary statement should always be used to make the intent clear to any unsuspecting inhabitants or people around.

Using mechanical and physical methods

BAIT BOXES. Bait boxes do not kill or trap rodents. They are simply used to introduce poison baits safely in the job site. However, their use for trapping by incorporating a snap trap or glue board inside is another supplementary way in which they can be utilized.

Bait boxes are designed as a tamper-proof box, so that a child or pet cannot get to the bait inside. Tamper-proof boxes vary in type and quality of construction, but they are usually made of heavy plastic. Rodent bait stations need to be secured to the floor, wall or ground. The following should be ensured while making use of them:

- Bait boxes should be clearly labelled with a precautionary statement.
- Bait boxes need to be checked periodically.
- Bait boxes should be placed wherever the rodents are most active as determined by droppings and other signs.
- Bait stations can also have an optional apparatus to provide drinking water. This can make the stations more acceptable, particularly when available water is scarce.
- It should be noted that rodents fear new objects at first, so baits may not be taken for the first few days or a week. Once bait is taken, the bait box needs to be left in place for some time. The rodents now consider it to be part of their normal surroundings. This is termed familiarizing.

USE OF TRACKING POWDERS. Rodents have an elaborate behaviour of grooming themselves by licking their fur. Tracking powder makes use of this behaviour. The formulation is a rodenticide carried on a talc that is applied in areas where rodents live and travel. The powder sticks to the rodents' feet and fur and is swallowed when the rodents groom themselves. The major advantage to tracking powder is

Table 9.2. Characteristics of commonly used rodenticides in pest control. (From http://npic.orst.edu/factsheets/rodenticides.html.)

Rodenticide	Type	Chemical class	Days of feeding needed
Brodifacoum	Anticoagulant	Hydroxycoumarin	Single
Bromadiolone	Anticoagulant	Hydroxycoumarin	Single
Bromethalin	Non-anticoagulant	Other	Single
Chlorophacinone	Anticoagulant	Indandione	Multiple
Cholecalciferol	Non-anticoagulant	Vitamin D3	Multiple
Difethialone	Anticoagulant	Hydroxycoumarin	Single
Diphacinone	Anticoagulant	Indandione	Multiple
Strychnine	Non-anticoagulant	Alkaloid	Single
Warfarin	Anticoagulant	Hydroxycoumarin	Multiple
Zinc phosphide	Non-anticoagulant	Other	Single

that it can kill rodents even when food and water are plentiful, or if rodents have become bait- or trap-shy.

- Reading the label directions before use is a must.
- Application should be with a hand bulb, bellows duster, or with a flour sieve or salt shaker.
- Tracking powders should never be used in suspended ceilings, around air ventilators, or near food or food preparation areas as the powder can become airborne and drift into non-target areas.

9.2 Bats

Bats often come into conflict with humans as they tend to find shelter in structures (Fig. 9.9). Bats are beneficial animals and participate in the pollination of a large number of trees, help in seed dispersal and also consume large numbers of insects.

9.2.1 Controlling and management of bats

Bat control should be planned as a long-term rather than a short-term solution. Often short-term solutions can be hazardous to both bats and humans. The following methods recommended by Pierce (2003) are most suitable to control and manage bats.

Inspecting for bats

The first step in a bat control programme is bat inspection. Solitary bats often enter a building through an open window or door while searching for food. Bats may be seen flying around a porch light, chasing the insects attracted to the light. At times large numbers of bats can find shelter in a structure. The

Fig. 9.9. Bats can be accidental intruders or a resident pest. (Courtesy of USDA Forest Service Southern Research Station, USDA Forest Service, SRS, Bugwood.org.)

inspection should point to all bat exit and entrance points, as well as the number of bats.

Using bird netting to exclude bats from a house

Bats do not gnaw or claw their way into a building as rats and mice will. Therefore, almost anything can be used as a temporary seal – fibreglass insulation, rags, oakum, steel wool, etc. Permanent bat proofing requires materials such as sheet metal, plastic netting, plywood, caulking compound or aerosol foam insulation.

Using bat repellents

When exclusion is not possible or feasible these products can be used to provide some level of control.

- *Naphthalene* (crystals or flakes) is the only chemical currently registered as a bat repellent for indoor use. Naphthalene should be applied at the rate of 2.3 kg per 57 m³ of attic or wall void space. As the material vaporizes, the bats will be repelled and will not return as long as the strong odour remains. Once the material dissipates, the bats will return. Humans should avoid breathing the fumes, and sensitive people or those with respiratory problems should avoid all treated areas.
- *High-frequency sound* in the range of 4000–18,000 cps has been used successfully to repel bats from gymnasiums, large warehouses and similar structures.
- *Bright lights* strung through an occupied attic to illuminate all roosting sites may repel bats. The key to success with this method is to make sure that all roost sites are illuminated. Large attics may require several 100- to 150-watt bulbs. This method is cleaner and safer than other methods.

Controlling the occasional bat

On occasion, one or two bats may get into a home and fly around. When this happens, there is no need to panic, for it will not attack even if it is chased. The bat usually will find its way back outdoors by following fresh air movements. Leave windows or doors open to help it escape. Also, turn off all lights: if any are left on, the bat may seek refuge behind wall hangings or drapes. If a bat refuses to leave, it can be caught with a net, coffee can or gloved hand and released outside.

9.3 Birds

Like many other animals, birds create problems when they come into conflict with humans. Bird droppings from perches over warehouses, malls and residences, as well as birds flying across airfields, are common complaints (Figs 9.10 and 9.11). When this happens, control measures – not necessarily lethal – should be taken to solve the problem. The presence of birds in the area or neighbourhood is not detrimental. In fact, birds provide many more benefits than most people realize.

Control measures must provide a long-term solution to the problem. Many short-term control measures do not work and may pose a hazard to both birds and humans. Scarecrows, gun shots, recordings of dying birds, spikes and other deterrent objects and other devices are available on the market and give the impression that they repel birds. They work pretty well, but only for a short while.

9.3.1 Control and management of birds

Bird spikes

These are spikes or nails that can be fixed on all areas of the building where birds perch frequently.

Nets

The area where the birds frequently perch can be covered by fine nets, preventing them from freely

Fig. 9.10. A dent in an aeroplane engine caused by a bird hit. (Courtesy of NAIA, Manila.)

Fig. 9.11. Birds killed by aeroplane propeller hits. (Courtesy of NAIA, Manila.)

perching. This deters the birds and usually makes them move out.

Bird repellent chemicals

Bird repellents are possibly the best solution to long-term bird management. These products are available as fogging solutions. One of the bird repellents is methyl-anthranilate (MA), an organic compound that has been tested and successfully used in area-wide bird management programmes. MA has been shown to function as a repellent by acting as an irritant on the bird's taste buds, skin and trigeminal chemoreceptors in the beaks, gizzards, eyes and mucous membranes.

Bird scaring gels

There are many kinds of repellents, but one unique one is a gel formulation with the property of reflecting light as a sparkle in the UV-A light spectrum, which has proven to be more effective than the rest. Birds see light in the UV-A spectrum, and to them the gel looks like a sparkle resembling flames. This virtual 'flame shield' visually deters the birds from landing. The gel's odorous herbal extract is an additional deterrent, which works in case the bird takes a chance and lands.

Use of birth control baits

Nicarbazin is a complex of two compounds (4,4′-dinitrocarbanilide (DNC) and 4,6-dimethyl-2-pyrimidinol (HDP)). At a certain dosage it has been shown to affect egg laying, interrupting egg laying and reducing hatchability in birds. Presently this compound on corn-based bait is being used for pigeon control in Europe (Fig. 9.12).

Fig. 9.12. Corn-based bird baits used as feed for pest pigeons. (Courtesy of David Loughlin.)

10 Methodology in Pest Control – Insecticide Formulations

10.1 Pesticide Formulations and Application

Pesticides have dramatically changed human lives in allowing increased food production, lowering risks from vector-borne diseases and keeping a check on various nuisance pests. The need and demand for pesticides is bound to grow with rapid urbanization and increased human intolerance.

Urban pesticides can be divided into three main classes in general as follows:

- Insecticides – chemicals used to kill insects.
- Herbicides – chemicals sprayed to kill or control weeds.
- Fungicides – chemicals used to kill or control fungus and moulds.
- Rodenticides – chemicals used to kill rodents.

All pesticide formulations are a mixture of one or more active ingredients, a synergist (for some formulations) and non-active ingredients (sometimes referred to as inert). An active ingredient is a substance that prevents, kills or repels a pest. There are various categories of active ingredients as shown in Table 10.1.

Synergists are a type of active ingredient that is sometimes added to formulations. They enhance another active ingredient's ability to kill the pest while using the minimum amount of active ingredient, but do not themselves possess pesticidal properties. For example, insecticides containing the active ingredient pyrethrin or pyrethroid often contain piperonyl butoxide (PBO) as a synergist.

Non-active ingredients are ingredients used to aid in the application of the active ingredient. These are solvents, carriers, adjuvants, dyes or any other compound intentionally added.

Insecticides are conventionally applied by mixing the formulation into water and by the use of various types of handheld sprayers such as calibrated compressed air sprayers (Figs 10.1 and 10.2), thermal fogging machines, mist blowers, low-volume (LV) sprayers, ultra low-volume (ULV) sprayers, etc. Also backpack and truck-mounted sprayers are commonly used for area-wide application.

10.2 Commonly Used Insecticide Active Ingredients

Generally insecticides used in urban pest control are categorized as shown in Table 10.1.

10.2.1 Pyrethroids in general

Pyrethroids are the most common active ingredients in the pest control industry now. They have high insecticidal potency and low mammalian toxicity and have found widespread acceptance after restrictions were imposed on the organophosphate and carbamate groups of chemicals. Acute and chronic toxicity are considered low with pyrethroids and thus they are more popular in the industry. Pyrethroids are used in making practically all types of formulations except baits. Most formulations with pyrethroids are also categorized as green label, giving them an advantage over the others in household use. For example, bifenthrin, deltamethrin, permethrin, lambda-cyhalothrin, alpha-cypermethrin, etc.

Mode of action

Pyrethroids act by keeping open the sodium channels in neuronal membranes affecting both the peripheral and central nervous systems. This leads to a hyper-excitable state causing such symptoms as tremors, incoordination, hyperactivity and paralysis. They are effective against most insect pests and are extremely toxic to fish.

10.2.2 Organophosphates – chlorpyrifos and dichlorvos

Chlorpyrifos is in the organophosphate (OP) group of compounds. In most Asian countries chlorpyrifos

Table 10.1. Common categories of insecticide active ingredients in use with general modes of action.

Pyrethroid	Organophosphates	Benzoylurea	Phenylpyrazole	Neonicotinoid	Diamide
Nerve poison	Nerve poison	Mostly growth regulators	Effect on nervous system	Effect on nervous system	Type of muscle poison

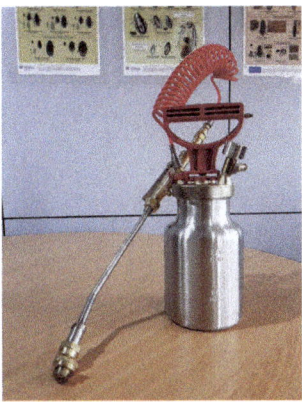

Fig. 10.1. (a) A typical handheld metallic compressed air sprayer. (Author's photo.)

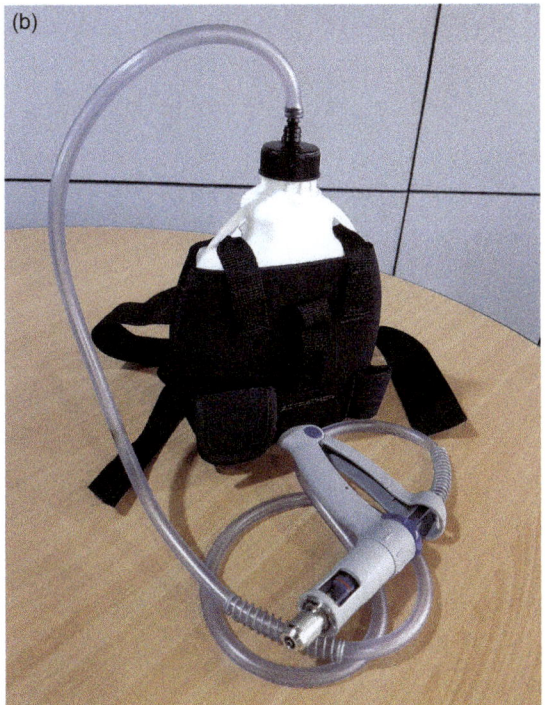

Fig 10.1. (b) A body-wearable plastic sprayer for specialized spraying. (Author's photo.)

is commonly used for termite protection as a barrier or soil poisoning chemical. Occasionally, it is applied directly on wooden surfaces against wood borers and drywood termites.

Residues and deposits from chlorpyrifos application can remain on household objects like rugs, furniture, stuffed toys and other absorbent surfaces. Concerns over its toxicity led the United States Environmental Protection Agency (EPA) to cancel and phase out nearly all residential usage of chlorpyrifos in the year 2000; however, it is still popular in other parts of the world.

Dichlorvos (DDVP) is in the OP group of compounds. It is extensively used in developing countries against household pests and is available in concentrates, strips and aerosols. Its usage has been restricted in most developed countries. DDVP vaporizes quickly and provides quick action on pests with lower residues. This vaporizing property of DDVP makes the product a good choice for cracks and crevices and surface treatment.

Mode of action of OPs

OPs cause acetylcholinesterase (AChE) inhibition and accumulation of acetylcholine at neuromuscular junctions, leading to paralysis and death. It has very broad-spectrum use and is highly toxic to aquatic organisms.

10.2.3   Insect growth regulators in general

Unlike classic insecticides, insect growth regulators (IGRs) do not affect an insect's nervous system, or act as a stomach poison, but inhibit growth of insects in various ways. The most popular IGRs are the chitin synthesis inhibitors (CSIs) and juvenile hormone (JH).

Mode of action

CSIs work primarily by preventing the formation of chitin, which is the building block of insect exoskeletons. With these inhibitors in action the moulting

Fig. 10.2. Different forms of liquid formulations (from left to right): emulsion concentrate (EC), Gel, suspension concentrate (SC) and microencapsulation (ME). (Author's picture.)

process is disrupted. This causes the insect to die, either quickly or after several days. For example, chlorfluazuron, diflubenzuron, hexaflumuron, noviluron.

JH-based IGRs typically work by mimicking or inhibiting the JH analogue, one of the two major hormones involved in insect moulting. IGRs that mimic JH can produce premature moulting of young immature stages, disrupting larval development and eventually killing the developing insect. For example, pyriproxyfen, methoprene, hydroprene.

10.2.4 Carbamates – propoxur

Propoxur is a methyl carbamate insecticide and is commonly used as a household insecticide. It is sold both as a consumer product and a professional product. Propoxur is mostly available as concentrate and in aerosol form for control of cockroaches and flies.

Mode of action

Carbamates cause AChE inhibition affecting the central nervous system (i.e. rapid twitching of voluntary muscles and, eventually, paralysis and death). It has very broad-spectrum use and is highly toxic to fish.

10.2.5 Phenylpyrazoles – fipronil

Fipronil is a broad-use insecticide that belongs to the phenylpyrazole chemical family. Fipronil is a widely used insecticide used to control urban pests including cockroaches, ants, termites, and to treat fleas and

ticks on pets. It is formulated as baits, sprays, dusts and aerosols for application. However, its most popular usage is in making insecticidal baits for cockroaches and ants. Fipronil serves as a good bait toxin not only because of its slow action, but also because it is non-repellent.

Mode of action

This group of compounds block the inhibiting receptors leading to an excitation of the nervous system. It leads to neuronal hyper-excitation due to accumulation of the neurotransmitter gamma-aminobutyric acid (GABA) at the synaptic junctions. Its mode of action is, therefore, antagonistic.

10.2.6 Neonicotinoids – imidacloprid

Imidacloprid is a neonicotinoid insecticide that produces neurotoxicity through fully binding or partial binding to specific areas of the nicotinic acetylcholine receptor. Imidacloprid is popularly used as a soil termiticide, in baits for control of ants, cockroaches and houseflies, and in pet products for fleas and ticks.

Mode of action

This compound works by interfering with the transmission of stimuli in the insect nervous system as it causes a blockage of the nicotinergic neuronal pathway. By blocking nicotinic acetylcholine receptors, this compound prevents acetylcholine from transmitting impulses between nerves, resulting in the insect's paralysis and eventual death.

10.3 Commonly Used Insecticide Formulations

There are a number of insecticide formulations that essentially serve the same purpose but differ in the composition of the carrier material (Fig. 10.2). Some of the major ones are discussed below.

10.3.1 Emulsifiable or emulsion concentrate (EC)

EC formulations usually contain a liquid active ingredient, one or more petroleum-based solvents and an agent that allows the formulation to be mixed with water to form an emulsion. Each litre of an EC formulation generally contains 5% (50 g/l) to 50% (500 g/l) active ingredient. ECs are among the most versatile formulations. They are adaptable to many types of application equipment, including small portable sprayers, hydraulic sprayers, LV ground sprayers and mist blowers (Table 10.2).

Table 10.2. Advantages and disadvantages of EC formulations.

Advantages	Disadvantages
Cheap and cost effective	Odorous
Little or no agitation required before mixing with water	May cause unwanted harm to plants and non-target areas
Relatively easy to handle, transport and store	Easily absorbed through the skin of humans or animals
Non-abrasive	Solvents may cause rubber or plastic hoses, gaskets, and pump parts and surfaces to deteriorate faster
Will not clog sprayer screens/filters or nozzles	May cause staining or discolouration on painted surfaces
Can be used in thermal foggers	Flammable (should be used and stored away from heat or open flames)
	May be corrosive

10.3.2 Suspension concentrate (SC)

SC formulations generally make use of solid active ingredients that are insoluble in water. These may be formulated as flowables in which the finely milled active ingredients are mixed with water, along with inert ingredients to form a fine suspension. These are then further mixed with water for application and are similar to EC or wettable powder formulations in ease of handling and use (Table 10.3).

Table 10.3. Advantages and disadvantages of suspension concentrate formulations.

Advantages	Disadvantages
No odour	Require some agitation before mixing with water
Mostly water based, safe for handling	May leave a visible residue
Easy to handle and apply	Occasionally clog finer nozzle types
Higher residual efficacy	Often expensive
No staining or discolouration on painted surfaces	

10.3.3 Microencapsulation (ME)

Microencapsulated formulations contain fine particles or droplets of the active ingredient surrounded by a coating to give additional function. These formulations are highly specialized and provide specific properties such as controlling the release of active ingredients, extending residual properties, reducing irritability of the active ingredient, reducing decomposition of the active ingredient, etc. The capsules are generally dispersed in a solvent such as water for easy miscibility and use.

10.3.4 Aerosols

Aerosol formulations are formulations packed in compressed form containing one or more active ingredients and a solvent as a propellant. Most aerosols contain a low percentage of active ingredients. Aerosols are either in a ready-to-use type, or as a smoke or fog generator. These formulations are best used for an on-the-spot quick action solution (Table 10.4).

Table 10.4. Advantages and disadvantages of aerosols.

Advantages	Disadvantages
Ready to use	Limited use as only a few pests can be addressed
Useful for deep penetration and quick flush out action	Risk of inhalation injury
Easily stored and carried	No residual action
Convenient way to buy small amount of formulated pesticide	Hazardous if punctured, overheated or used near an open flame
Retain potency over fairly long time	

10.3.5 Granules

Granular formulations contain granular particles, which are larger and heavier than dust formulations. The coarse carrier particles are made from an absorptive material such as clay, sand or nut shells. The active ingredient is either coated on the outside of the granules or is absorbed into them. Granular pesticides are most often used as baits, for control of larval mosquitoes and for slow-release requirements (Table 10.5).

Table 10.5. Advantages and disadvantages of granules.

Advantages	Disadvantages
Ready to use, no mixing	Do not stick to surfaces
Drift hazard is low, and particles settle quickly	Need an attractant to draw the pest into contact with the granule
Little hazard to applicator (no spray, little dust)	May need moisture to start pesticidal action
Simple application equipment, such as seeders or fertilizer spreaders	May be hazardous to non-target species once left in open
Break down more slowly through a slow-release coating	

10.3.6 Dusts

Dust formulations are always in a ready-to-use form and contain a low percentage of active ingredient. The active ingredient is milled with a very fine dry inert carrier such as talc, chalk, clay, nut hulls or volcanic ash. Dusts are always used in moistureless conditions. Dusts are used to control termites, lice, fleas and other parasites on pets and livestock (Table 10.6).

Table 10.6. Advantages and disadvantages of dusts.

Advantages	Disadvantages
Usually ready to use, with no mixing	Only for specific pests
Effective where moisture from a spray might cause damage	Easily drift off-target during application
Require simple equipment	Residue easily moved off-target by air movement
Have longer residual effect	May irritate eyes, nose, throat and skin
	Does not stick to surfaces

10.3.7 Fumigants

Fumigants are formulations that are in the form of poisonous gases once applied. The advantage of fumigants is that they are liquids or volatile liquid or solid when packaged under controlled conditions, but change to gases when they are released. Other active ingredients are volatile liquids when enclosed in an ordinary container under pressure. Fumigants are used in food and grain storage facilities in greenhouses, granaries and grain bins, as well as for whole-house treatment against multiple pests (Table 10.7).

10.3.8 Baits

A bait formulation is a mixture consisting of an active ingredient, food and an attractant. The bait either attracts the pests or is placed where the pests will find it. Pests are killed by consumption of the pesticide contained in the bait. The amount of active ingredient in most bait formulations is very low, usually less than 2%. Indoors, they are used to control cockroaches, ants, houseflies, termites and rats. Outdoors they are used to control snails, slugs, birds and rodents (Table 10.8).

Table 10.7. Advantages and disadvantages of fumigants.

Advantages	Disadvantages
Toxic to a wide range of pests	The target site must be enclosed or covered totally to prevent the gas from escaping
Can penetrate tightly packed areas such as soil or grains under certain conditions	Highly toxic to humans and all other living organisms
Single treatment usually will kill most pests in treated area	Require the use of specialized protective equipment, including respirators and trained personnel
Time saving and economical	Require the use of specialized application equipment
Minimal residues	No residual action
	Urban use is highly restricted or limited

Table 10.8. Advantages and disadvantages of baits.

Advantages	Disadvantages
Ready to use	Slow action as it takes more time to affect the individual pest as well as the entire group or colony
Targets specific to intended pests	Often expensive
Non-toxic as it uses less toxin	Requires bait stations, follow-up monitoring and bait replenishment
Allows spot treatment and less intensive compared to spraying	Need skill to apply
Controls pests that are cryptic in nature and difficult to reach	
Allows monitoring, so job can be evaluated to give high efficacy	

11 Methodology in Pest Control – Insecticide Baits and Baiting

11.1 Baits and Methods of Baiting

Baits and baiting procedure have many advantages over conventional pesticide sprays. The advantages include minimal impact on humans who are living or working amid infestations (Table 11.1).

11.2 Baiting Cockroaches

Baiting cockroaches using food-based baits is a proven strategy (Fig. 11.1). However, the work requires some degree of skill and training.

Cockroaches are not social insects and lack the recruitment behaviour shown by termites and ants. Each cockroach makes its own decisions; this makes the task of baiting more challenging. The performance of a bait or baiting process thus depends on integration of knowledge of cockroach feeding preferences, patterns of movement, resting behaviour and the nature of the job site. In addition, the fact that German cockroaches, for example, are not attracted to bait over long distances makes bait performance dependent on proper placement of bait.

- *German cockroaches are not attracted to bait over long distances, thus bait performance is dependent on proper placement of bait.*

There is evidence that poor quality baits as well as improper baiting that fails to kill in its first or subsequent encounters provide opportunities for the cockroaches to learn to avoid future contact. This learned avoidance of bait can be overcome by choosing a good bait and also by periodically moving the baits to new locations and also rotating bait with different matrices and active ingredients.

Cockroaches are known to associate visual cues with food and use learned cues to forage. Usually a new food type placed in a new site attracts more cockroaches than the known food type in the known site. When the known food type was in a new site and the new food type in the known feeding site, most of the cockroaches oriented towards the known food type and neglected the new one. These results revealed that many factors influence the discovery and ingestion of a food source and that cockroaches make different foraging decisions in relation to the existing environmental situation. They are able to distinguish a novel food placed in a novel site from a novel food placed in a site previously occupied by another food type. This also means that cockroaches learn the location of specific food resources and associate particular locations with particular resources. This associated learning behaviour is useful in optimizing bait application.

- *Failure to determine harbourages and applying bait far from harbourages renders bait ineffective. Bait placed a few feet away sometimes has no or minimal effect.*

It is well known that German cockroaches are reluctant to venture far from areas of harbourage for foraging and this makes the role of inspection in the process of baiting even more important. Failure to determine harbourages and applying bait far from harbourages renders bait ineffective. Bait placed a few feet away sometimes has no effect. The bait placement guides included with most bait products are important to optimize bait performance.

Table 11.1. Practitioner's perspective on baits and baiting method.

Baiting	Practitioner's perspective
Handling	No odour, no site preparation, does not require mixing
Safety	Lower exposure to pesticides, lower chances of contamination and spills
Type of job site	Any site, particularly sensitive sites can be well treated
Time taken to treat	No mixing and less preparation prior to application
Application nature	Precise, spot treatment. Allows monitoring and follow up so the treatment can be adjusted and modified
Pest coverage	Multiple pests such as cockroaches, ants, flies, termites, rodents are addressed
Economics	Cost–benefit favourable over long term

The success of bait and baiting programmes depends on the following:

- quality of the bait gel;
- technical skills of the bait applicator;
- pest population;
- harbourage location;
- sanitation of the area;
- bait placement; and
- follow-ups and monitoring.

11.2.1 Common active ingredients used in cockroach bait

There are a number of active ingredients used in cockroach control baits. A few of them are described here.

- *Fipronil*: Fipronil-based baits may be the fastest acting bait on the market as it is non-repellent and readily transfers to other members of the group. Fipronil is effective at very low concentrations. It is available in granular and gel formulations.
- *Hydramethylnon*: Hydramethylnon is a slow-acting stomach poison. It is low in toxicity to mammals and birds. It is mostly available as a granular bait, and as gel in syringe applicators.
- *Abamectin*: Abamectin is a toxic extract from a soil microorganism with low toxicity to mammals. It comes in a gel bait formulation.
- *Indoxacarb*: Indoxacarb is considered a reduced risk insecticide because of its low toxicity to animals. Once eaten by the cockroach, it is converted into a chemical toxin inside the cockroach body. It readily transfers from one cockroach to others through the horizontal transfer effect, increasing its effectiveness. It is available as a gel bait form,

and as a flowable dust to be applied in cracks and crevices.
- *Imidacloprid*: Imidacloprid is readily soluble in water and has a very low odour. The toxic effects are highly specific against cockroaches and extremely low towards vertebrates. It is available as a gel bait for cockroaches in a syringe applicator.

11.2.2 Baiting cockroaches with a sprayable form of bait

One of the most advanced technologies in cockroach control is the use of microencapsulated formulations to 'attract and kill' the pest (Fig. 11.2). The technology uses microcapsules to hold the active ingredient on both porous and non-porous surfaces for much longer than conventional residual formulations. Effective and prolonged control is ensured by the mixture of two types of capsules – one containing a food grade, the attractant, which diffuses out from the capsule attracting the pest to a treated site. The other carries a toxicant that slowly diffuses out from the capsule at a controlled concentration and kills the pest on contact.

The product is applied with a conventional sprayer as spot treatment in and around harbourage sites. The spray attracts the cockroaches to come out and walk over the treated surface and they are then killed (Figs 11.3 and 11.4). Cockroaches not killed due to lower exposure carry some of the microcapsules to their harbourage, in turn passing over poison to group mates. These secondary exposed individuals are eventually killed over time.

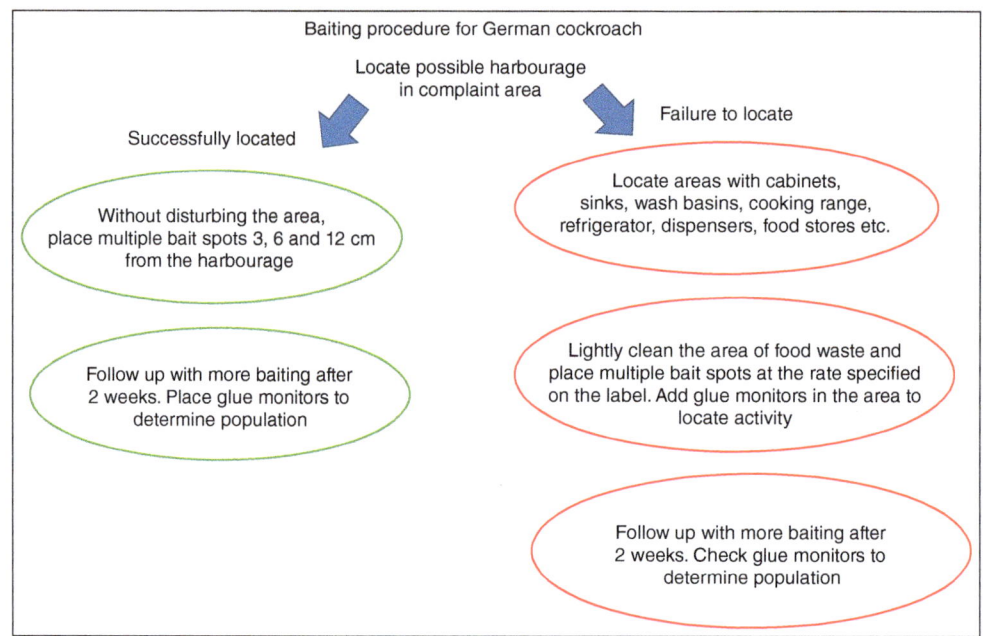

Fig. 11.1. A simplified baiting procedure for German cockroach. (Author's illustration.)

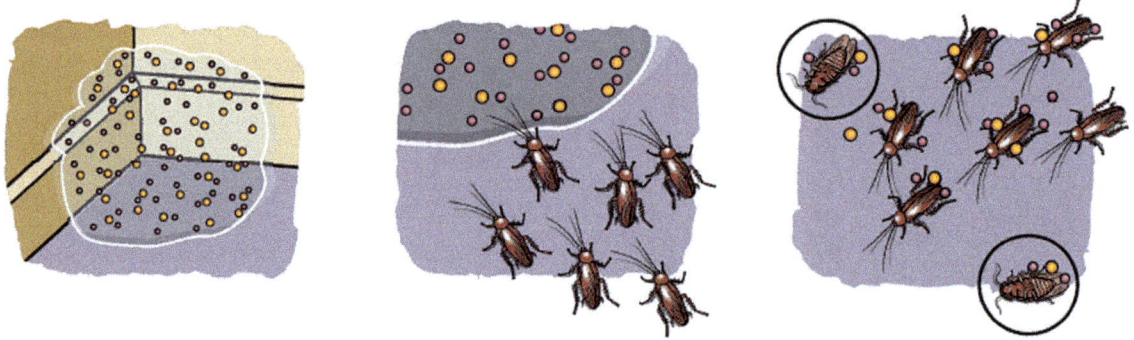

Fig. 11.2. Mode of action of sprayable bait. Cockroaches contact the attractive bait capsules and trample them which releases the active ingredient, thus killing them. (Courtesy of ICB Pharma.)

11.3 Baiting Common Ants

Ants are opportunistic feeders and will take whatever foods are available. But in general they are divided into sugar and fat feeders. Most quality commercial baits available for ants are complete mixtures and are readily acceptable to both types of pest ants. However, it is always advisable to identify the ants before taking up baiting.

Ant baits on the market are available in both solid granule and liquid gel form (Fig. 11.5). It is also known that exposure of ant bait to sun, shade,

moisture and dry conditions produces differences in efficacy results irrespective of the active ingredients used. Both gel baits and granular baits can also differ in their efficacy with ageing and exposure of the bait in storage and service.

11.4 Baiting Subterranean Termites

Termite baiting is applicable only for species belonging to the subterranean termite group. The procedure allows above-ground infestation to be

Fig. 11.3. Actual footage of aggregating German cockroaches when sprayed with bait. (Courtesy of Josielyn Trinidad.)

Fig. 11.4. Spraying bait as spots near cockroach harbourage. (Courtesy of ICB Pharma.)

instantly treated with the long-term effect of colony elimination.

- *The major advantage of baiting over chemical treatment is that baiting kills the colony and pre-vents re-infestation and consequently eliminates the need for frequent retreatment.*

Termite baiting is a sure process to eliminate sub-terranean termite colonies. It is used for both above-ground and in-ground infestations and is thus applicable for undertaking both preventive and corrective types of treatment.

Another advantage of termite bait is that the method helps reduce the amount of active ingredient

Fig. 11.5. Granular ant baits used for controlling indoor pest ants. (Author's photo).

in use. An example is shown below that compares recommended dosages as per label of various active ingredients for termite treatment currently in use in the Philippines (Table 11.2). The values are computed based on the commercial label and market research. Baiting requires minimal use of the active ingredient compared to other types of treatment.

11.4.1 Preventive treatment using in-ground stations

This method is used as a preventive measure for termite treatment (Figs 11.8–11.11). The structures treated with this method remain under protective monitoring. In-ground stations are installed in the ground, around the perimeter of the structure. These stations are then monitored periodically (every 4–6 weeks depending upon the geography). Once a termite is detected in the station/s, bait is added and the colony is eliminated.

An in-ground station consists of:

- a perforated cylindrical plastic station with a top lid which helps aggregate the termites; and
- wooden interceptors which help keep the termites in the station.

Installation procedure

In-ground stations are placed in the ground around the exterior of the structure at intervals of 3–5 m.

Table 11.2. Comparison of the amount of active ingredient used between soil treatment chemicals and baiting.

Active ingredient	Method of use	Minimum quantity of active ingredient required to treat a 100 m² house per application (in grams)
Chlorpyrifos	Soil injection	1920
Imidacloprid	Soil injection	200
Fipronil	Soil injection	117.6
Bifenthrin	Soil injection	150.4
Chlorfluzuron	Baiting	1.5–2.3

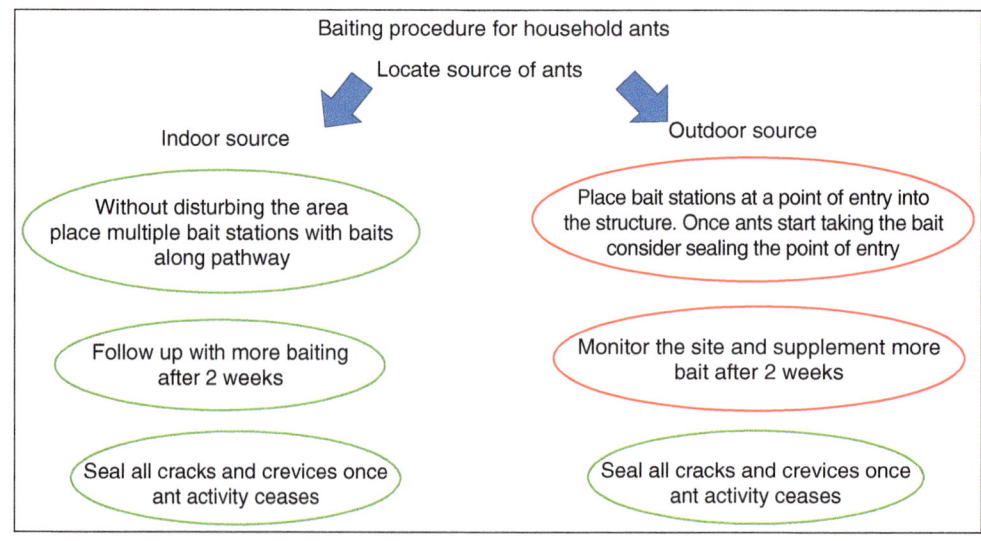

Fig. 11.6. A simplified procedure for baiting common indoor ants. (Author's illustration.)

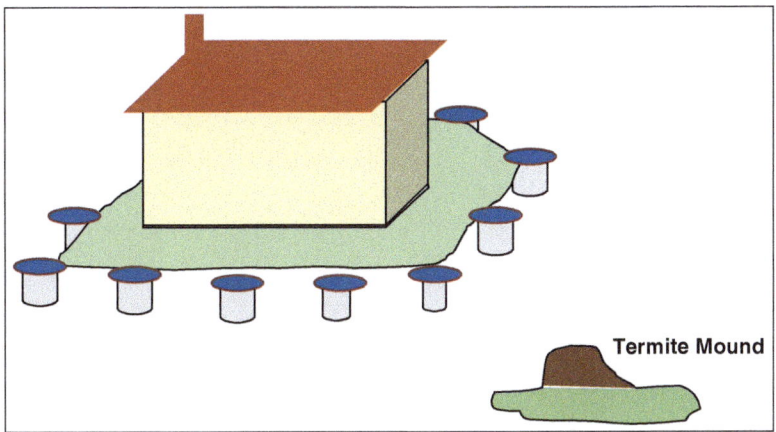

Fig. 11.7. In-ground stations are installed around the perimeter of the structure as a system to provide a continuous monitoring service.

Fig. 11.8. Use of a handheld augur to make a hole for the in-ground station installation. (Author's photo.)

The distance between stations and the structure should be ideally as close as possible, preferably within 0.5–1.0 m (Fig. 11.7). However, stations placed nearer or further will also work.

If soil is unavailable it is necessary to form holes in slabs or asphalt and place concrete stations (ICSs) (Figs 11.12–11.14).

Precautions

- Care should be taken to check the stations are not installed in a disturbed area, such as

pathways, drainage, buried electric lines, low lying flooding or water runoff zones, etc.
- Care should be taken to bury the stations fully in the ground with only the lid above the ground. Sunlight penetrating inside stations will deter termites from being intercepted.

Station inspection

In-ground and in-concrete stations are inspected regularly for evidence of termite activity. Usually the inspections are done every 4–6 weeks. Inspections

Fig. 11.9. Installing an in-ground bait station. (Courtesy of Michael Home.)

Fig. 11.10. A typical in-ground station with wooden interceptors. (Author's photo.)

Fig. 11.11. Closing the in-ground station with a secure lid. (Author's photo.)

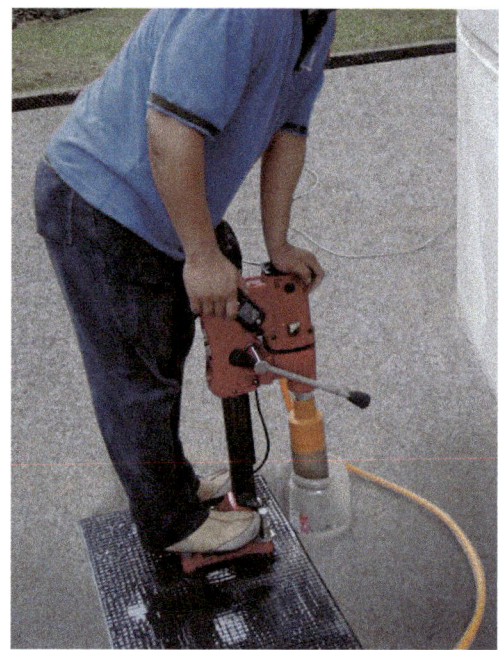

Fig. 11.12. Drilling a hole through the slab to put in an in-concrete station. (Courtesy of Zen Mathias.)

Fig. 11.13. An in-concrete station in place. (Author's photo.)

are repeated for a year or as the agreement specifies. Signs of mud tubes or live termites feeding on the wooden interceptors inside the stations are noted visually. When such termite activity is discovered, termite bait is prepared as per product specification and label directions and then added to the station. Termite bait is added only to the stations that are noticeably active with termites. Other stations are left for routine monitoring.

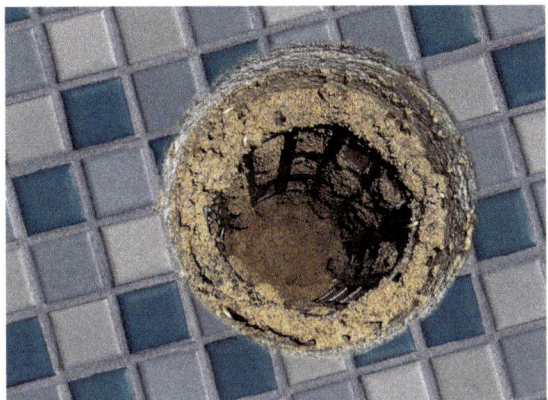

Fig. 11.14. An in-concrete station intercepting a termite colony. (Author's photo.)

Fig. 11.15. (a) Baiting an in-ground station with processed cellulosic termite bait. (Author's photo.)

Baiting

Bait preparation should be based on a careful reading of the instructions and as per the instructions given out during product training (Figs 11.15a and b). This step is extremely important as termites are very sensitive to disturbance, quality of food presentation, moisture and contamination.

Station re-inspection and re-baiting

Baited stations are re-inspected every 4–6 weeks. Stations are continually refilled with bait in order to make sure that adequate bait is provided to the colony. Total depletion of the bait in a station prior to colony elimination may result in the colony abandoning the station. In the event of such a situation it would take some time for the termites to get back to feeding. This delays the process of colony elimination.

11.4.2 Treating infestation with above-ground stations

Active infestations above the ground are treated with above-ground stations (Figs 11.16a and b). Areas where termites are visibly active feeding anywhere in the structure are the best spot to start using the stations. Installing stations in damage spots with no visible termites may not work in colony elimination. Termites are known to abandon infesting areas once they determine there is insecurity, disturbance or lack of food.

Fig 11.15. (b) Baiting an in-ground station with a form of compressed tableted bait. (Author's photo.)

An above-ground station consists of:

- a rectangular perforated plastic station with a top lid which helps feed the termites.

Installation procedure

Above-ground stations are placed directly on top of the infested site with visible termite activity. Stations should not be installed in damage sites without termite activity. Installation is done by using metal screws or construction grade tape. Once secure the station is filled with bait and then closed with the lid. Termites pass through the bottom perforations to reach the bait.

Station re-inspection and re-baiting

Stations are inspected every 4–6 weeks and bait consumption is monitored. New bait is added if consumption continues until stoppage. Four weeks from complete stoppage of feeding, the stations are removed and the damaged area is restored.

Precautions

- Live termites need to be detected before installing an above-ground station.
- The installation has to be done with minimum disturbance.
- Securing the station to the spot using screws and tape is a must.

Fig. 11.16. (a) An above-ground station installed on an indoor infested site. (Author's photo.)

Fig. 11.16. (b) An above-ground station installed on an outdoor infested site. (Courtesy of Paul Yuen.)

- Care should be taken not to install the station in areas of any type of disturbance such as rights-of-way.
- A precautionary notice should be written on the stations, such as 'Do not disturb'.

11.4.3 How to measure the process of colony elimination

Termite colony elimination is a slow process as it takes time to poison and affect all the members in the colony. The number of individuals in the colony may range from 100,000 to 2,000,000. The bait takes time to reach the majority of the individuals through the process of inter-feeding or tropholaxis. Thus, patience is a virtue when working with termite baits.

The following are some measurable steps to determine that the colony is undergoing elimination. In the absence of any of these noticeable pointers it should be reasonable to accept that the bait is not working.

1. Consumption of bait – continuous and slowed down over time.
2. Physiological changes – dark pigmentation/blackening on the abdomens of the workers (Fig. 11.17).
3. Behavioural changes – sluggish movement and not avoidant of light when exposed.
4. Changes to termite caste ratios – more soldier than worker castes in the station (Fig. 11.18).
5. Reduced numbers – dead termites and less visible termites in the station.
6. No more termites – finally, no live termites in the station, stoppage in feeding.

Fig. 11.17. An excavated colony showing dark pigmentation on workers' bodies. (Author's photo.)

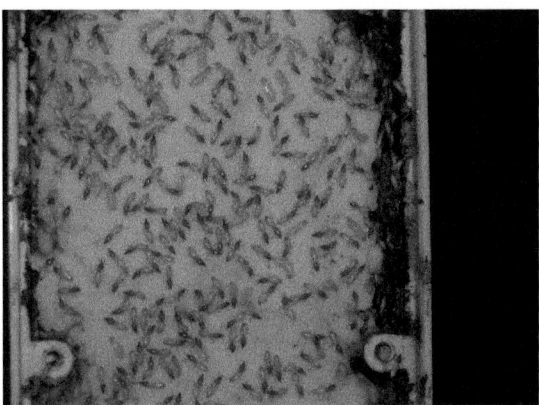

Fig. 11.18. A bait station showing bait consumption and change of caste ratio from workers to soldiers in the baiting station. (Author's photo.)

Active ingredients commonly used in termite bait

Chitin synthesis inhibitors (CSIs) are currently the only active ingredients that have been shown to cause colony elimination in termites.

CSIs are very powerful against termites, and have low toxicity to humans and other non-target organisms such as mammals, birds and reptiles. The most common active ingredients used in making termite baits are as shown in Table 11.3.

Table 11.3. Common active ingredients used in making termite baits.

Active ingredient	Composition of bait
Bistrifluron	10.0 g/kg bait
Chlorfluazuron	1.0 g/kg bait
Diflubezuron	2.5 g/kg bait
Hexaflumuron	5.0 g/kg bait
Noviflumuron	5.0 g/kg bait
Novaluron	5.0 g/kg bait

12 Shift to Integrated Pest Management (IPM)

Integrated pest management (IPM) is a methodology practised by pest control practitioners who employ human judgement in their work. The reason for the necessary shift to IPM from conventional pest control activity is that over-reliance on pesticides has led to repeated control failures. In addition it is felt that the use of pesticides is making pest control more and more difficult, as all forms of pesticide applications are coming under greater scrutiny by clients. The only solution to this growing concern is the adoption and advocacy of IPM as the primary approach to solve pest problems.

A planned incorporation of various control methods into a pest management programme is defined as IPM. IPM goes well beyond regular and scheduled use of pesticides. It not only requires understanding of pest biology, ecology and behaviour, but also diverse knowledge on buildings and structures, functioning of the structure, the occupants' life styles, landscaping work and types of intervention methods available. However, in spite of proven successes, practitioners have shown limited interest in adopting IPM. The primary reasons could be the perception that higher investments of money and time are required for IPM than for conventional pest control. However, numerous reports exist in which IPM is shown to be cost competitive and more effective than conventional treatments. These studies have proven IPM to be viable, workable, ecological and profitable to pest control practitioners. Overall IPM has been proven to be sustainable.

12.1 IPM Versus Conventional Pest Management

The key benefits of conventional methods of pest control are that they can cover a wider range of pests, provide quick and easy elimination and have long field persistence. Conventional methods depend on the use of pesticides as a 'stand-alone' approach to pest control in which the chemical provides significant or an acceptable level of reduction in the pest population. It involves a single action of chemical application following a regular, predetermined spray schedule.

IPM works on the principle that pest control is more than just eliminating pests. More importance is attached to maintaining control of pests and preventing re-infestations. IPM programmes have a number of key attributes to offer when it comes to maintaining control. Each IPM programme follows stringent monitoring and intervention methods to keep a check on the pest population. This usually involves a combination of numerous methods such as monitoring tools, barriers, chemicals and education.

12.2 IPM has to be Marketed as a Value-Added Service

The success of IPM depends on making it distinguishable from conventional pest control by features that would eventually appeal to consumers. A number of such attributes could be used as value additions to the service. This would allow the consumer to appreciate and invest in IPM programmes. The following points could be value additions for IPM programmes, which will help distinguish IPM from conventional pest control.

- *Success of IPM depends on making it distinguishable from conventional pest control using special features that would attract consumers.*

12.2.1 Providing a diagnostic service

Identifying the source of the problem and successive investigations are critical in determining the actual pest problem. Tools, devices and aids need to be involved in correctly diagnosing the site before making a proposal. Correctly evaluating the situation will help pinpoint the area that needs to be treated, the strategy to be used, the type and amount of insecticide to be used and much more. This is a value addition to the service job that the client will take note of.

12.2.2   Installing pest barriers

IPM heavily relies on physical methods, and barriers are very useful. Pest barriers are permanent fixtures to a structure preventing entry of pests. These could be sealants, wire mesh, metal plates, special cement or any other intervention that becomes part of the structure. Non-resident pests frequently invade structures through even very small entry points: adult mice can pass through an opening as small as 1 cm wide and cockroaches need only a 2.5 cm space to enter a building. Once pests get inside, forcing them out is much more challenging and often requires the use of chemicals. To keep pests outside, practitioners can identify and seal potential pest entry points. Certain pest barriers could even make a structure more energy efficient by sealing up gaps.

- *Pest barriers are physical items and this visible installation appeals to consumers. It gives them a sense of reassurance as the act is undertaken to improve the quality of the structure.*

Such barriers could also help control elements such as heat, light and moisture in the structure, which are add-on advantages. Pest control proposals that include provision for sealing gaps and fixing cracks and crevices for pest prevention, as well as eliminating the pest itself, are often found by customers to be more appealing and justifiable than conventional pest control work.

12.2.3   Use of safe and efficient formulations

IPM programmes provide avenues to choose the right type of formulation to be used and move away from conventional stand-alone types. This would eventually help reduce the amount of pesticide used and make IPM a safer choice. It would also help consumers distinguish IPM programmes from conventional services.

The nature of the pesticide formulation to a great extent dictates the pest control strategy. Sprayers have helped to deliver pesticide with relative ease, but also have allowed its indiscriminate usage, often leading to contamination of non-target areas. This is a concern when using a sprayer within the confines of a building. In recent times we have seen the development of a number of high performance formulations for use in the pest control industry. Formulations such as microencapsulation, capsule suspension, dry-flowable, gels, granules and baits are some examples that are popular among practitioners. The advent of baits for a variety of pests has further reduced regular insecticide sprays and introduced the necessity for inspection and monitoring as key features of IPM.

12.2.4   Use of baits and colony suppression

Conventional insecticides are often not effective in managing structural pests such as cockroaches, ants, termites and rats, which are cryptic in nature. Also when disturbed they move out of the treated zone. Insecticide baits have provided a solution for such hard-to-control pests by enabling targeted treatment to inaccessible areas. The process of 'horizontal transfer' of insecticide across to other members of the group or colony has been successfully elucidated in cockroaches, termites and ants by the use of baits. This function of the bait formulation is most helpful in eliminating pests that live in groups or colonies.

12.2.5   Providing options and warranties

Knowing consumers have poor knowledge of pest management, practitioners can address the 'information asymmetry' and make the business more profitable by offering extra benefits to gain consumers' trust. Information asymmetry between consumer and practitioner can be further addressed by providing literature, client references and web reviews on both products and quality of services. Practitioners can use product updates with research and findings to help consumers gain reassuring information in order to aid the decision-making process. Practitioners can also experiment with offers of property damage replacement warranties, which are known to work. These efforts will help practitioners to persuade consumers to invest more in pest management and, in turn, this will allow the practice of IPM.

Another way of overcoming consumer fear is by offering a long-term IPM programme, such as an annual program instead of a single treatment. This is more reassuring to consumers. The practitioner may include an additional benefit by including another service, such as sealing pest entry points or monitoring cockroaches in the annual programme. The probability of the consumer agreeing to purchase such a package is greater according to my current survey in the Philippines. Consumers are frequently offered free cockroach or rodent control services along with annual termite control and maintenance programmes. Studies have shown that when facing possible risk in a deal, consumers tend to build in considerable margin for error – what might be gained must be worth a good bit more to compensate for the unpredictability. By offering additional services without an extra charge in the above scenario, the pest controller creates an environment where the consumer gets a chance for some compensation and this in turn helps with the decision-making process.

12.2.6 Providing a multi-service programme

It is understood that managing pests using IPM is often perceived as expensive. In an industry where price remains pivotal, the success rates for adoption of IPM will thus depend on the cost being justified. Practitioners are thus challenged to develop methods and skills to present IPM in a manner that will make it acceptable to consumers. Practitioners could present additional service benefits and incorporate them into their programme. Two successful case studies by the author are reported in this section to show that presentation of IPM as a multi-service programme can help its acceptance and adoption over conventional treatments in spite of the higher cost.

● *Practitioners are challenged to develop methods and skills to present IPM in a manner that will make it acceptable to consumers. One way is to present additional service benefits and incorporate them into their programme.*

Case study no. 1

A case study in the Philippines is presented here using termite management as an example. It has been accepted that successful termite management relies on continuous monitoring and maintenance. Both monitoring and maintenance in turn make termite management programmes expensive. A 6 ha private beach resort was used as a test ground for evaluating a methodology for presenting IPM to the property owners. All the 17 structures in the resort showed termite infestation. It was also learnt that for over 10 years the structures were treated by periodic soil and direct spot spray treatment. To convince the

resort owners to shift to a comprehensive IPM programme a proposal for a total change in paradigm was prepared and presented.

The proposal presented to them was a 5-year project, which involved a number of distinct jobs, namely:

1. The control of existing termites in the structures.
2. Regular monitoring of the structures.
3. The interception and eradication of incoming termite colonies.
4. The control of termites in landscape and avenue trees in and around the structures.
5. Sanitation, involving the removal of unprotected wood and wooden objects.
6. The destruction of visible termite mounds.
7. A reduction in the amount of toxic chemicals used.
8. Documentation and presentation of a regular progress report.
9. The conducting of seminars for in-house maintenance staff.

The costs for each of the services were separately computed and presented in a total. In spite of the cost being a few times higher than the conventional treatment method, the owners decided in favour of going ahead with the IPM programme. This success helped justify the conclusion that there is the need for practitioners to project the number of value-added services in IPM that justify its higher cost and make each venture profitable and sustainable.

Case study no. 2

A mosquito control programme for a gated community consisting of 262 homes spread across a 3.7 ha property was used to prove that presentation of IPM as a multi-service programme helps its acceptance and adoption over conventional treatment. The programme consisted of a number of distinct value-added services such as:

1. Install CO_2 mosquito traps.
2. Provide regular mosquito population counts.
3. Identification of mosquitoes.
4. Survey water bodies within the property for mosquito breeding.
5. Eliminate breeding sites.
6. Conducting space treatments.
7. Collecting feedback through a text message system.
8. Notifying home owners of the times of outdoor mosquito activity.

9. Submit report with copies emailed to individual home owners.

10. Conduct seminars on mosquitoes and mosquito control along with local officials.

The above IPM programme replaced a conventional pest control programme, which involved bimonthly fogging of the community with a monthly treatment of drains, manholes and canals with granular larvicide. The cost of the IPM programme was much higher than the conventional programme, but as the programme involved a number of value-added services, it was accepted by the home association and was implemented successfully.

12.2.7   Presentation of cost–benefit overview

A creditable evaluation of IPM programmes requires an improved theoretical understanding of the relationship between IPM and conventional pest control inputs. Improper understanding or failure to recognize the cumulative benefits leads to the notion that IPM ventures are expensive. In fact in the long term all IPM ventures I have experienced or come across have proven to be successful both on economic and environmental grounds. A study by Brenner (2003) reported the cost of an individually tailored IPM was equal to or lower than traditional chemically based pest control. In another study by Sever *et al.* (2007) IPM was shown to cost less than conventional treatment. Some studies have shown that an easily replicable single IPM visit was more effective than the regular application of pesticides in managing cockroach populations (Kass *et al.*, 2009). These findings demonstrate that customized IPM can be successful and cost-effective in an urban community, and also that there is a strong possibility of a client choosing a multiple task-based IPM programme over a conventional programme despite any extra expense. The extra cost is justified by the usage of fewer chemicals and minimizing exposure of the inhabitants in addition to controlling the pest successfully.

Poor popularity and adoption levels of IPM are due to a failure of practitioners to understand and promote the key attributes of IPM. Offering the key attributes of IPM would allow consumers to pay a higher rate and in return get satisfactory service. Practitioners also fail to educate consumers on the added values of an IPM programme.

- *Value addition will help consumers distinguish IPM from conventional pest control and eventually lead them to select it as the method of choice. The key value addition of IPM that would appeal to consumers most is the reduced levels of active ingredient used.*

12.3   Shifting to IPM

The key value addition of IPM that would appeal to consumers most is the reduced levels of chemicals used. This specific attribute continues to be the most attractive factor that appeals to consumers as shown in surveys. As urban pest control takes place in close association with the human population, the chance of potential human exposure to pesticidal products is high. IPM would reduce this exposure by incorporating a number of calculated actions including replacing old formulation types with new ones. Newer formulations have safer active ingredients and delivery methods, which in turn cuts down the chances of human exposure. In this regard the development of insecticide baits has helped IPM considerably in not only making the application inspection driven but also more environmentally friendly. However, in spite of baits being safe, their application requires higher skill levels and takes longer than other methods, which makes them less popular with practitioners. It is not uncommon to see that the practitioner's choice of formulation and the nature of the pest control programme are determined by cost and ease of application rather than logic, which remains a challenge.

13 Handling Pesticides

Pesticides are toxic to humans, animals, birds and fish. However when handled carefully by trained practitioners as per label directions they do not present any hazard. *Incorrect handling, such as accidental spillage on body parts, inhalation or consumption can pose harmful effects including fatality.* Such cases should be immediately handled in consultation with the local poison centre or a physician.

The most common pesticide exposure is through inhalation, which is often overlooked. This happens when a pesticide bottle is opened, and the chemical is being poured, mixed and sprayed. *Most new-generation pesticides do not have a smell or are termed odourless, preventing their detection by nose. Pesticide vapours from these odourless formulations are being inhaled without being detected.* A complete face mask or a mask over the nose, mouth and eyes is always advisable when handling any type of pesticide.

13.1 Using a Pesticide

There are four important things a practitioner should do before handling pesticide.

1. Read the pesticide label.
2. Understand the hazard class.
3. Know the hazard pictogram.
4. Keep emergency measures on hand.

13.1.1 Reading the pesticide label

All pesticide packaging has a label (Fig. 13.1). The purpose of the pesticide label is to protect human health and the environment. In Asian countries, pesticide labels are legal documents in that they are required by law to be put on pesticide packaging. Labels serve as the only contact between the manufacturer and the user of the product. They convey essential usage recommendations and safety information. In addition, the label informs the user about the hazards of the pesticide, and risks of its use, which should help the user to assess the actual risk of handling and applying the product under specific local conditions.

13.1.2 Understanding the hazard class

The World Health Organization (WHO) has introduced a colour band scheme, indicating (mainly) acute health hazards of the pesticide product, which is based on the WHO classification of pesticides by hazard. *In this scheme, bands of specific colours indicating the hazard classification of the product are printed horizontally on the bottom part of the label.*

Practitioners are thus required to check the colour and use the product accordingly. Table 13.1 shows the colour classifications as per the WHO directive.

13.1.3 Knowing the hazard symbol pictograms

Hazard symbol pictograms (or hazard symbols) are those pictograms visualizing the hazard (or sometimes risk) of the product, according to the Global Harmonizing System (GHS) (Fig. 13.2). The following hazard symbols are defined by the GHS. They should be in the shape of a square set at a point, and should have a black symbol on a white background with a red frame. The exact size of the hazard symbol pictogram will depend on the size of the pesticide container/label, but must not be less than 10 × 10 mm in size.

13.1.4 Keeping emergency measures on hand

Poisoning due to pesticides is usually acute and results from extensive skin contact or ingestion. Signs and symptoms vary with the type of pesticide and can sometimes be confused with other illnesses. If any of the following symptoms are detected a

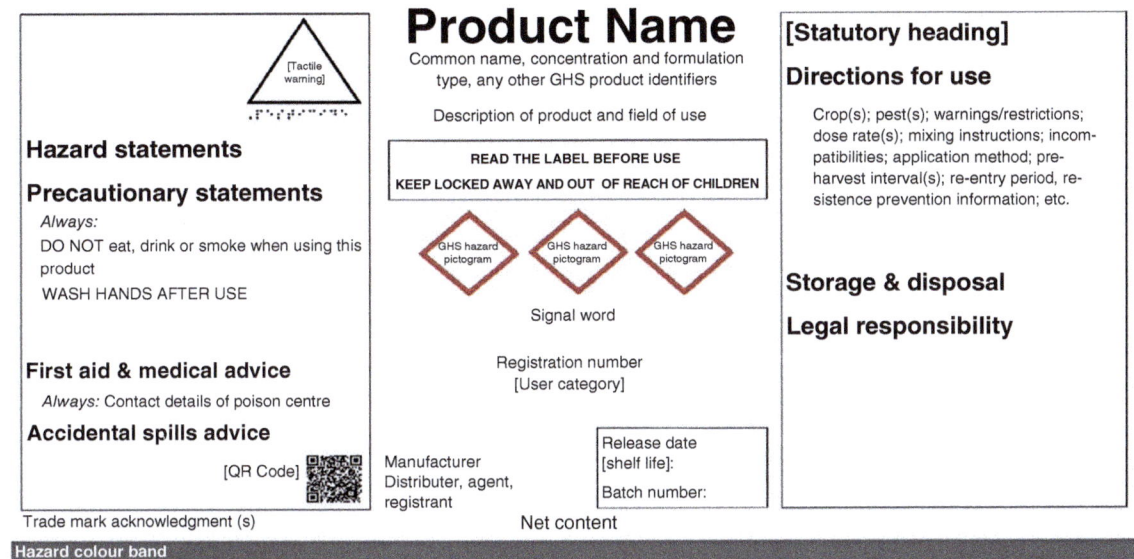

Fig. 13.1. A typical label explaining complete details of the product, its method of use, storage and first aid procedures. (Food and Agricultural Organization, 2015.)

Table 13.1. WHO classified toxicity indications are used in all product labels to aid user friendly understanding. (Food and Agricultural Organization, 2015.)

	Hazard class				
Class	Class Ia	Class Ib	Class II	Class III	Class U
Degree of hazard	Extremely hazardous	Highly hazardous	Moderately hazardous	Slightly hazardous	Unlikely to present any hazard in normal use
Hazard symbol	☠	☠	✗	No symbol	No symbol
Signal word	Very toxic	Toxic	Harmful	Caution	No signal word
Colour band	(red)		(yellow)	(blue)	(green)

consultation with a qualified physician and immediate treatment should follow.

Indications of pesticide poisoning:

- General: extreme weakness and fatigue.
- Skin: irritation, burning sensation, excessive sweating, staining.
- Eyes: itching, burning sensation, watering, difficult or blurred vision, narrowed or widened pupils.
- Digestive system: burning sensation in mouth and throat, excessive salivation, nausea, vomiting, abdominal pain, diarrhoea.
- Nervous system: headaches, dizziness, confusion, restlessness, muscle twitching, staggering gait, slurred speech, fits, unconsciousness.
- Respiratory system: cough, chest pain and tightness, difficulty with breathing, wheezing.

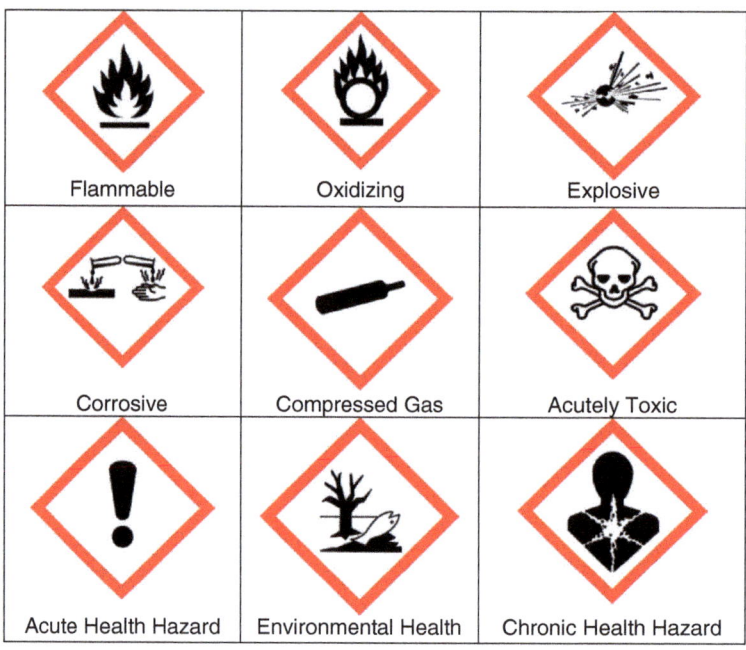

Fig. 13.2. Globally harmonized pictograms for easy product categorization and classification. (Food and Agricultural Organization, 2015.)

Appendix

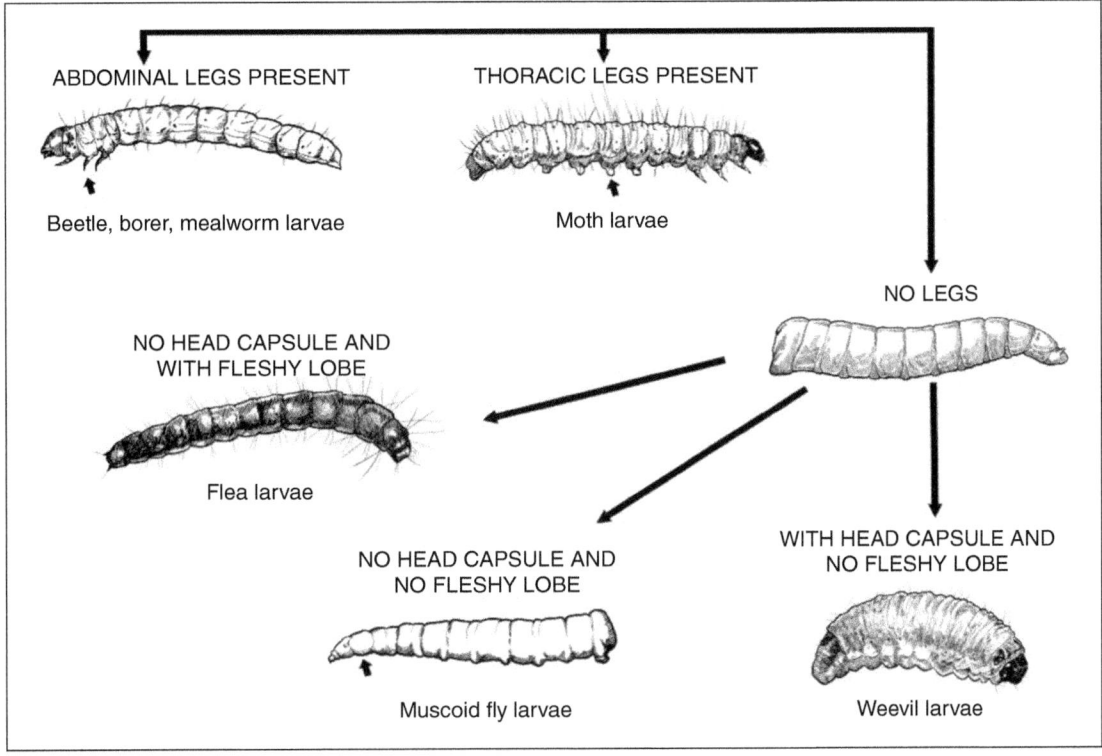

Fig. A1.1. Simple identification key to common household pest larvae.

	American cockroach
	German cockroach
	Brown banded cockroach
	Oriental cockroach
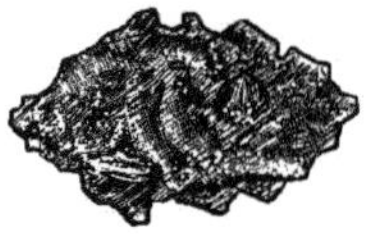	Australian cockroach

Fig. A1.2. Simple identification key for cockroach droppings. The size of the droppings generally vary from 0.5-3.0 mm in length.

Table A1.1. Toxicities of some commonly used active ingredients on the urban pest control market. The values indicated are for rats in mg/kg body weight.

Active ingredient	Major use category	LD_{50} oral	LD_{50} dermal
Abamectin	Insecticide	300	>1800
Alpha cypermethrin	General insecticide	400–700	>2000
Azadirachtin	Growth regulator	>5000	>2000
Bacillus thuringiensis	Mosquito larvicide	>5000	
Beta cyfluthrin	Insecticide	70–650	>5000
Bifenthrin	General insecticide	262	>2000
Bistrifluron	Chitin synthesis inhibitor	>5000	
Chlorpyrifos	General insecticide	92–276	2000
Chlorfluazuron	Chitin synthesis inhibitor	>8500	
Cyfluthrin	General insecticide	500	>5000
Cyhalothrin-lambda	General insecticide	79	632
Cypermethrin	General insecticide	250	2000
Cyromazine	Growth regulator	3387	>3100
Deltamethrin	General insecticide	431	>2000
Diazinon	General insecticide	300–400	3600
Diflubenzuron	Insect growth regulator	4640	>10,000
Emamectin	General insecticide	1516	>2000
Fenoxycarb	Growth regulator	16,800	>2000
Fipronil	General insecticide	336	382
Hexaflumuron	Chitin synthesis inhibitor	>5000	
Hydroprene	Growth regulator	>34,000	5100
Imidacloprid	General insecticide	450	>5000
Indoxacarb	Growth regulator	268	>5000
Malathion	General insecticide	5500	>2000
Novaluron	Chitin synthesis inhibitor	>5000	–
Piperonyl butoxide (PBO)	Synergist	>7500	–
Permethrin	General insecticide	>4000	>4000
Pyrethrin	General insecticide	1500	>1800
Pyriproxyfen	Growth regulator	>5000	>2000
Pyrethrum (BO)	General insecticide	1500	>1800
Rotenone	General insecticide	132–1500	–
S-methoprene	Growth regulator	>34,000	>2000
Spinosad	Mosquito larvicide	>5000	
Tetramethrin	General insecticide	>5000	>2000
Thiamethoxam	General insecticide	>5000	>2000

Table A1.2. The World Health Organization (WHO) now uses the Acute Toxicity Hazard Categories for classification. These are as follows.

WHO class	LD_{50} for the rat (mg/kg body weight)	
	Oral	Dermal
Ia Extremely hazardous	<5	<50
Ib Highly hazardous	5–50	50–200
II Moderately hazardous	50–2000	200–2000
III Slightly hazardous	Over 2000	Over 2000
U Unlikely to present acute hazard	5000 or higher	

References

Brenner, B.L., Markowitz, S., Rivera, M., Romero, H., Weeks, M., Sanchez, E., Deych, E., Garq, A., Godbold, J., Wolff, M.S. *et al.* (2003) Integrated pest management in an urban community: a successful partnership for prevention. *Environmental Health Perspectives* 111, 1649–1653.

Center for Disease Control (2013) Leishmaniasis FAQs. Available at: https://www.cdc.gov/parasites/leishmaniasis/gen_info/faqs.html (accessed 15 October 2017).

Dhang, P. (2011) An attempt to termite-proof structures using physical barrier in the Philippines. In: Forschler, B.T. (ed.) Proceedings of the 10th Pacific-Termite Research Group Conference, Hanoi, pp. S4.

Dhang, P. (2014) *Urban Insect Pests: Sustainable Management Strategies*. CAB International, Wallingford, UK.

Diclaro, J.W., Cohnstaedt, L.W., Pereira, R.M., Allan, S.A. and Koehler, P.G. (2012) Behavioral and physiological response of *Musca domestica* to colored visual targets. Neurobiology, physiology, biochemistry. *Journal of Medical Entomology* 49(1), 94–100.

Food and Agricultural Organization (2015) *International Code of Conduct on Pesticide Management Guidelines on Good Labelling Practice for Pesticides*. FAO, Rome.

GISD (2017) Global Invasive Species Database. Available at: www.iucngisd.org/gisd/species.php?sc=960 (accessed 15 October 2017).

Hogsette, J.A. (2008) House fly (Diptera: Muscidae) ultraviolet light traps: design affects attraction and capture. Proceedings of the Sixth International Conference on Urban Pests, Budapest, pp. 193–196.

Kass, D., McKelvey, W., Carlton, E., Hernandez, M., Chew, G., Nagle, S., Garfinkel, R., Clarke, B., Tiven, B., Espino, C. and Evans, D. (2009) Effectiveness of an integrated pest management intervention in controlling cockroaches, mice, and allergens in New York City public housing. *Environmental Health Perspectives* 117, 1219–1225.

Koehler, P.G., Oi, F.M. and Andrews, C.A. (2013) *Powderpost Beetles and Other Wood-Infesting Insects*. The Institute of Food and Agricultural Sciences (IFAS), Florida, USA. Available at: http://edis.ifas.ufl.edu/pdffiles/IG/IG11900.pdf (accessed 15 December 2017).

Pierce, R.A. (2003) Bats of Missouri: information for homeowners, University of Missouri. Available at: http://extension.missouri.edu/p/G9460 (accessed 15 October 2017).

Potter, M.F. (1996) Parasitic mites of humans. University of Kentucky. Available at: https://entomology.ca.uky.edu/ef637 (accessed 15 October 2017).

Rozendaal, J.A. (1997) *Vector Control – Methods for Use by Individuals and Communities*. World Health Organization, Geneva, Switzerland.

Sever, M.L., Arbes Jr, S.A., Gore, J.C., Santangelo, R.G., Vaughn, B., Mitchell, H., Schal, C. and Zeldin, D.C. (2007) Cockroach allergen reduction by cockroach control alone in low-income, urban homes: a randomized control trial. *Journal of Allergy and Clinical Immunology* 120, 849–855.

Sliney, D., Gilbert, D. and Lyon, T. (2016) Ultraviolet safety assessments of insect light traps. *Journal of Occupational and Environmental Health* 16, 413–442.

Further Reading

Bonnefoy, X., Kampen, H. and Sweeney, K. (eds) (2008) *Public Health Significance of Urban Pests*. World Health Organization Regional Office for Europe, Copenhagen.

Dhang, P. (ed.) (2014) *Urban Insect Pests: Sustainable Management Strategies*. CAB International, Wallingford, UK.

O'Connor-Marer, P. (2006) *Residential, Industrial, and Institutional Pest Control*. University of California, Oakland, California.

Robinson, W. (2005) *Handbook of Urban Insects and Arachnids*. Cambridge University Press, Cambridge, UK.

Index

Note: Page numbers in **bold** type refer to **figures**
Page numbers in *italic* type refer to *tables*